Techniques for Preparing and Obtaining Your Own Patent

by

Hrand M. Muncheryan

Howard W. Sams & Co., Inc.
4300 WEST 62ND ST. INDIANAPOLIS, INDIANA 46268 USA

FIRST EDITION
FOURTH PRINTING—1979

International Standard Book Number: 0-672-20995-0
Library of Congress Catalog Card Number: 73-83367

Printed in the United States of America.

Preface

Since there is an increasing trend of invention by laymen, this book was written to explain the procedures involved in filing for a patent. In addition to the necessary expenses of model making and application fees, the layman faces the increasing cost of professional help in application preparation. To help the layman defray some of these costs, detailed instructions have been included to enable the reader to prepare his own application.

It must be noted that a person filing his own application must be well versed in the English language in order that the Patent Examiner, or any layman, can fully understand the description of the invention as written by the applicant. When the application is not written in clear and concise terms, there may be misunderstanding between the examiner and the applicant, although the examiners are very lenient and helpful to the applicant who is seriously and honestly pursuing the prosecution of his application. All correspondence with the Patent Office must be carried out tactfully and with decorum.

There have been frequent changes in the United States Patent Rules during the past few years and the latest rules of patent practice and patent fees are incorporated in this book. New material on drafting drawings and preparation of comprehensive, valid claims is also given. An appendix is inserted at the back of the book to give examples of the utility patent, design patent, and a trade mark. It should be further noted here that this book, when followed according to the logical steps as presented, will assist the inventor in securing a patent when no prior patents or disclosures are cited by the Examiner during the prosecution of the application. For interference, infringements, and other litigations that may come up on random occasions, the inventor must resort to the services of a qualified professional patent attorney, licensed to prosecute patent cases in court.

Hrand M. Muncheryan

Contents

Foreword

Everyone has contact every day with a hundred things that either need to be invented or improved to a more efficient degree. It has been said that the greatest room in the world is the room for improvement, and nothing seems to be more to the point. When all is said and done, nothing is perfect. Many of the things we have today will have to be invented over again. For instance, take the present-day safety pin. It is simple and inexpensive and has brought a fortune to its creator. On first thought, no one would think that any improvement would be possible on this pin. Yet, only very recently, a patent was issued with three claims on an improved safety pin.

It is not necessary to be a genius to find out how to do something that has been done for the first time. It takes a critical attitude, which may be developed by any man. He must learn to look at a device and ask, "How can this be improved? How can it be made simpler and more cheaply?" Some engineers will find some ideas which they will feel sure have been patented before. It must be remembered that previous patents may be limited in scope or even useless and that the problem still remains for a more clever solution.

The field of inventing is vast and wide open to everyone interested in the subject. There are many areas where new ideas may be developed. Some examples in the area of defense might include: a method to determine an enemy plane from a friendly one; a means for deflecting a guided missile; countermeasures for homing missiles; an efficient method of protecting industrial areas against enemy attack. New types of transportation are possible on land, with the use of nitrogen or compressed air, and on the sea, with an invisible floatation device attached to the body. The laser beam is open to new applications for communications, at a minimum of five miles through fog and water, and in tracking moving objects, at low cost. Inventions in the field of entertainment might include a three-dimensional image television set which would operate without optical instruments worn before the eyes. A simple device for eliminating collisions between ships or aircraft would be most welcome. Other inventions might be: a practical method of refrigeration using, perhaps, the well-known Peltier effect; a new and effective source of producing light, video images, and memory devices; transmutation of matter on laboratory scale without danger to the worker. Obviously, the possibilities are endless. New ideas and pioneering have made America great and will continue to make it great—they are American heritage.

Many persons solve problems, but only few can discover them. It is they who have discovered, spent many an hour of creative thinking, and challenged the problem to its final solution that can achieve success with consequent personal satisfaction. But, it is rare to find both inventive ability and business knowledge in one person. Success, at times, depends on team work. Working for an employer who can offer the team and the necessary facilities gives you the opportunity to rediscover your hidden talents and to expand them at your will and the employer's cost.

Most any average person likes to have his capabilities and achievements recognized; this is only natural. There is no better or more appropriate place to demonstrate the importance and merits of your ideas than with your own company; you gain recognition for yourself and for your employer. No greater challenge to mental creativeness has ever existed in any period in history than it does today. It's your move next.

1

History of the Patent

Prior to the 15th century, an inventor was lucky if he avoided being burned as a wizard. At a later period, although inventions were not liable to the same objections as before, politically powerful guilds or craft unions could have a law passed providing a severe penalty for the user of the invention. However, in the course of time, it was realized that although some inventions might harm some particular individuals, they make life easier for the people as a whole. Consequently, a logical method of handling the situation was to encourage the inventor by granting him a monopoly for his invention.

In 1624, the British Parliament passed the Statute of Monopolies, which restricted the granting of monopolies for anything other than *new* inventions. The British Statute of Monopolies remains the basis of the British and American patent systems to this day.

In the United States, patent laws were enacted by Congress on April 10, 1790, in accordance with Article I, Section 8, of the Constitution. The Patent Office as a distinct bureau was established in 1802 under a Superintendent of Patents. Later, the Secretary of State alone was made responsible. In 1836, Congress made revisions in the patent laws and set up the Patent Office under a Commissioner of Patents, who performs all administrative duties with respect to the granting and issuing of patents and the registration of trade marks. Other revisions were made in 1870 and thereafter. The law now in effect in Title 37 of the United States Code is a general revision which was enacted on July 19, 1952, and went into effect on January 1, 1953. The Patent Office now is under the jurisdiction of the Department of Commerce.

WHAT IS A PATENT?

A patent for an invention is a grant by the government to the inventor, his heir, or assigns, of certain exclusive rights to his invention throughout the United States and its territories and possessions for the term of 17 years. It is a contract between the government and the inventor, giving the inventor, or his assign, the monopoly to prevent anyone else from making, using, or selling the invention. A distinction must be made here in that a patent does not grant the right to make, use, or sell the invention, but only grants the exclusive nature of the right. Any person is ordinarily free to make, use, or sell anything he pleases and a grant from the government is not necessary. The patentee, or his assign, may manufacture and sell the invention himself or he may license it; that is, he may give authorization to others to make and sell the invention.

In the United States, it is not the first person to file a patent application who is granted the patent. It is the first person who has reduced the idea to practice that is granted a patent, upon due showing and in compliance with the patent laws and regulations. Reducing an idea to practice means to make a drawing of, build, or produce a workable model which illustrates the invention. Proper drawings witnessed by two persons or by a notary public will constitute a valid step toward reduction to practice.

FUNCTIONS OF THE PATENT OFFICE

The principal function of the Patent Office is to administer the patent laws by granting letters patents for inventions. The Commissioner of Patents is the chief administrative head, and he has assistant commissioners under him. One of these assistant commissioners is in charge of the patent examining group, which consists of 15 different branches of examiners; each branch is headed by a chief examiner. The examiners examine the patent applications under their generic category and determine whether a patent should be granted, after making comparative studies of the application and the prior art. The examiner will notify the inventor regarding the allowability or rejection of the application.

PATENT PROSECUTION

Since the application for a patent is prepared in writing, the prosecution of the patent must also be in writing and must always be directed to the Commissioner of Patents, although it is the examiner assigned to the application who receives the correspondence and makes replies. An applicant for a patent may prosecute his own patent case, provided he has an adequate English vocabulary to describe his invention properly, and knows what his claims are to make his invention new, novel, and useful, over existing analogous inventions. A person lacking the ability to fully communicate his ideas and claims should not attempt to write his own application or prosecute it. He should get assistance from technical writers, patent agents, or patent attorneys and file his application himself, if he has to save some cost in the application. This book is not recommended for the latter type of applicant.

The following publications are recommended to the applicant:

(1) *Rules of Practice in Patent Cases.*
(2) *Patents and Inventions.*
(3) *Guide for Patent Draftsmen.*

They can be obtained by writing to the Superintendent of Documents, U. S. Printing Office, Washington, DC. 20402. Cost of each may change from time to time: therefore, request information before sending any money.

2

Types of Patentable Inventions

WHAT IS A PATENTABLE INVENTION?

There is no simple definition for a patentable invention. However, the requisites for patentability of a product are that it must be invented, must be novel, and must be useful.

In order to acquire a true understanding of the legal meaning of invention, various cases of the question based on actual court decisions are given below:

Q. Are minor improvements within range of expected skill patentable?
A. Generally no, unless such improvements are the only kind possible.
Q. Is superior workmanship, enlarging or strengthening of a machine patentable?
A. Absolutely not.
Q. May a function be patented?
A. No, but the means by which the function is performed may be subject of a patent.
Q. Is omission of a part or parts patentable?
A. Yes. Getting the same result with fewer parts may be a patentable invention of high order.
Q. Is duplication of parts in a machine patentable?
A. Not unless a new mode of operation results.
Q. Is it patentable to apply an old thing to a new use?
A. No, if the new use is analogous (a double use). Yes, if the new use is nonanalogous.
Q. Is adding an old part to a new combination patentable?

A. Not unless it produces a product of distinct character, different from that previously performed.

Q. Is the combination of old processes or materials patentable?

A. This depends upon the results. If the combination is not just an aggregation but involves an arrangement of elementary parts in successive cooperation to contribute to one, new, practical result, patentability is indicated.

Q. Is a theory of law of nature patentable?

A. No. Means or apparatus for making use of a theory or law of nature may be patentable.

Q. Is a new process for making an old thing patentable?

A. Yes.

Q. Does complexity of a device enhance patentability?

A. No. Mere complication is not invention.

Q. What is, in general, a patentable invention?

A. A new product may be subject of a patentable invention if it falls within the scope of any one of the following categories:

1. Involves a new mode of construction or operation.
2. Yields new properties or uses.
3. Produces a new function or increased efficiency.
4. Results in greater durability and reduced cost of operation.

TYPES OF PATENTS

The patent statutes distinguish three types of patents. They are utility patents, design patents, and plant patents. Only the first two will be discussed here because of their pertinence to machine industry.

Utility Patents

A utility patent involves the invention or discovery of any novel and useful art, machine, manufacture, or composition of matter. The word *art* here means a process, mode, or method; this is quite different from the common conception of artistry. Strictly speaking, a *method* involves mechanical operation, as in drawing a cup from sheet metal, whereas a *process* involves chemical changes in the material used. A patentable process is a method of treating certain materials to produce a particular predetermined result or product, such as heat treating steel to impart hardness to it, or treating cellulose in cuproammonium to produce rayon.

The word *machine* includes a part or a combination of mechanical parts designed to receive energy and to produce some useful end result. Tools and instruments operated by hand or some other source of energy are machines.

The word *manufacture* refers to a manufactured article, such as a valve, photographic camera, amplifier, ball-point pen, railway beam, and even an overhead crossing structure for highway intersections.

A *composition of matter* is a mixture of different materials which are combined to produce a substance having certain desired characteristics. Mechanical mixtures or chemical unions, whether solids, liquids, or gases, constitute composition of matter and may be the subject of a patent if they present invention, novelty, and utility.

Useless articles (manufactures) are not patentable, and the utility is negatived if the function performed by an article is injurious. This does not apply to war materials used by the Government. The validity of a patent on a useful article cannot be negatived for want of utility. Patents are never invalidated on the basis that the products covered by them may be imperfect in operation or crudely made. If an invention has any utility whatsoever, it is useful according to the patent statute. Of course a useful thing must also be novel, having a character different from anything that preceded it. Remember that a thing must be invented, must be novel, and must be useful in order to be the subject of a patentable invention. Utility patents are granted for a period of 17 years.

Design Patents

In making a reference to a patent, it is generally assumed that the patent is for an invention having utility (utility patent). A special form, known as a design patent, is granted to any person who has invented a new, original, and ornamental design for an article of manufacture. Inventions of design emphasize attractiveness of appearance rather than utility. In patent language, the word *design* is used in the artistic sense, not in the engineering sense.

There is a trend for incorporating certain ornamental features in an otherwise unpatentable article of manufacture in order to obtain a design patent and then to mark the article "Patented." While certain monopoly is enjoyed by the inventor, the general public is victimized by this practice, being led to the belief that the patentee has a utility patent. For this reason, the Patent Office is exercising strict rules in granting a design patent on anything whose principal value arises from its utility.

Articles protected only by a design patent should be marked "Design Patent" followed by a design patent number. Infringement rules for design patents apply in much the same way as in utility patents.

Design patents are granted for periods of 3½, 7, and 14 years, as desired and requested by the applicant. The design patent fees are paid according to the patent period. A fee of $20 should accompany each design application. When a notification is sent to the applicant that his application is allowable, he should then decide how many years of patent coverage he wants. The fee for 3½ years will be $10; for 7 years, $20; and for 14 years, $30. Or, the applicant may apply for any one of the periods desired during filing of his application by including the additional fee with the filing fee.

How to Determine Which Patent to Apply For

A patentable invention may be the subject of a utility patent, design patent, or both. For instance, a clock operating by energy from a minute radioactive source, if such function is novel, may be the subject of a utility patent. But if the housing of the clock contains artistic figures, such as nymphs, a star, a lion, or any artistic curve or relief not utilized in any prior patented art, it may also be the subject of a design patent. If both utility and design patents are desired on this clock, two separate applications should be filed. A design patent alone on this clock will not prevent anyone from making and selling the same clock without the design feature.

Since artistic tastes change with different periods, the design of a patented invention may be modified and new design patents obtained during the life of the utility patent and thereafter, if desired. Under infrequent, special conditions, and when an invention is the only basic type in existence, a patent may be reissued by a new application and upon payment of due fees.

3

Types of Granted Patents

PIONEER PATENTS

Patents that are granted on inventions when an art is young are pioneer patents. Edison's incandescent lamp, Land's Polaroid material, Farnsworth's electron multiplier, Zworykin's iconoscope, and Robertshaw's thermal controls may be classified as pioneer inventions.

BASIC PATENTS

Patents that are broad enough in scope to dominate a large part of an art are basic patents. While most pioneer patents may be considered as basic if broadly covered in a patent, a pioneer patent may be drawn too narrowly to be considered a basic patent. On the other hand, a basic patent may come too long after the art has been known to the public to be considered pioneer. Patents on items that are generally known by the names of their originators or trade names are basic. Such items are characterized by Gillette razor, Rayon, Wien oscillator, Hall process of refining aluminum, etc. Later patents on similar things are improvement patents.

IMPROVEMENT PATENTS

Patents in a given art subsequent to pioneer or basic inventions are improvement patents. These patents are those granted as an art grows older and more crowded. The flexible term art may be divided into larger arts and smaller arts. A patent in a larger art may be an improvement patent, and in a smaller art it may be a basic or pioneer patent, when finer improvements and discoveries are made on the subdivision of the larger art. For instance, in the larger art of hydraulic valves, a particular type of sealing gland may be a subject of a basic patent. The majority of the patents that are granted are improvement patents, as will be noted in the preamble of a patent application.

UNDERLYING PATENTS

Underlying patents are those pioneer and basic patents that have preceded certain improvement patents in an art. This is regardless of whether the pioneer or basic patents are still in force or have expired. Patents on Gillette, Rayon, Hall process, Polaroid, etc., are considered underlying patents.

DOMINATING PATENTS

When a patent covers an invention so fully that improvements on it or combinations involving it cannot legally be practiced without the consent of the patent owner, the patent is called a dominating patent. A dominating patent does not have to be a basic or pioneer patent. Any improvement patent may cover such a scope that it dominates any further improvement inventions, making any later improvement patents, if granted, subject of invalidation.

SUBSERVIENT PATENTS

Inventions whose making, using, and selling are controlled by a patentee by virtue of his granted patent are dominated inventions. Patents on such inventions dominated by broader patents are called dominated or subservient patents. Many subservient patents may be considered valueless.

ANTICIPATING PATENTS

Patents on inventions whose description and inventive matter are revealed in other publications or patents, so as to prevent the granting of subsequent valid patents, are known as anticipating patents.

UNITARY INVENTION PATENT

Inventions covered by utility patents are either unitary or combination. In a unitary invention, the useful result is obtained by a single unit—for example, a razor blade or a novel clip for holding papers together. The patent on such an invention is called a unitary patent.

COMBINATION PATENT

A combination patent is granted on an invention which requires more than one unit to produce a result and has claims reading on such a plurality of units. A combination patent may be granted on a particular kind of pen point and a particular type of pen-point holder. In such case, it would not be an infringement to make, use, or sell either the pen point or the holder if one or both are not protected separately by unitary patents. As was discussed in previous paragraphs, a combination patent must produce a unitary result. If more than one result is produced, such as a pencil with an eraser tip (each provides a separate function) the invention would be an aggregation and no patent will be granted.

FORTIFYING PATENTS

After a patent is granted on an article, minor changes may be made in the article, and these changes may be protected by an additional patent or patents for the sole purpose of strengthening the patent position. Such patents are called fortifying patents. Fortifying patents are useful in the prevention of circumvention of the patent by others.

NUISANCE PATENTS

Sometimes patents are obtained by an inventor for the sole purpose of annoying or hampering another. A manufacturer may buy a patent from an inventor to shelve it, thereby preventing his competitor from acquiring the patent and working it. Such patents are known as nuisance patents and are rarely put into practice.

4

Interpretation of Patents

MAKING A PATENT SEARCH

Prior to filing an application, the Patent Office will not advise an inventor as to the novelty of an invention or act as counselor for individuals. All records pertaining to issued patents are open to the public. When it is not convenient for an inventor to inspect these records, he may employ an attorney to make a patent search for his invention. However, the records of pending applications are preserved in secrecy and no information can be obtained on such cases until issued. At times, information may be exposed during an interference but only to the rival applicants.

In the search room of the Patent Office, the patents are arranged in classes and subclasses according to their nature. Most of the prior patents relating to a given invention may be found during the search. The value of such a search depends on the ability of the searcher and the time he spends on it. In litigation cases, months are sometimes spent in examining the files in search of prior art. However, such intensive and prolonged methods are not usually resorted to before filing an application. A simple search through proper classification should be sufficient, however inconclusive it may be.

It is usually impossible to advise that a patent can be obtained on an invention as a result of a patent search which has disclosed no anticipation. Nor can one state positively that an existing patent is valid, because there might exist uncovered prior patents or publications appearing in periodicals or books. Opinions of patent examiners as to what does or does not constitute invention also vary widely. In important patent cases, a validity search is made to determine whether any of the requisites of patentability were lacking in the prior patent when it was granted. When such is the case, the patent was mistakenly granted and is invalid. A favorable finding as a result of a validity search is usually considered a safe ground for avoiding infringement of that patent.

ANTICIPATION

A patent can be considered in three ways: (1) as a disclosure of a step in the evolution of the art, (2) as a possible anticipation of a future invention, and (3) as a claim on which one must not infringe. A prior patent anticipates an invention if it shows the invention in the drawing or specification. The disclosure in the specification of an application may show something physically different from the alleged anticipating patent, but unless the claim or claims of the application define this difference the claims are anticipated. Hence, one claim which fails to define such a difference may be anticipated, while another claim in the application which defines such a difference is not anticipated.

Sometimes a prior patent may suggest or predict that certain results may be obtained with the invention. However, the mere suggestion or prediction without sufficient information to enable one skilled in the art to make and use the invention claimed is not an anticipation. This point should always be considered in studying a patent during search, since it may mislead the reader as if it is an anticipation. A claim in an application may literally read on a prior patent or disclosure and yet not be an anticipation. If the item disclosed in the prior patent is incapable of performing the function defined by the claim in that prior application, the second application could not be an anticipation either. Sometimes a prior patent is found which seems to disclose both the structure and function defined by the claim of the application, but it will not anticipate the claim. This is because such a situation may be incidental, since its true nature and utility was unknown during the preparation of the application of the prior patent.

In summary, if there exists tangible evidence in the form of a prior patent, printed publication, or working model that the invention in question has been invented before, the prior device is said to anticipate the later application. In such an event, the Patent Office usually denies the grant of a patent on such application. On the other hand, if the Patent Office is not aware of the prior invention and grants a patent, this patent is liable to be declared void by a court during a law suit on this patent. Thus, it is discouraging to think that a granted patent may turn out to be an invalid patent. Such cases are not frequent, but they do happen.

DOMINATION

An invention is dominated by a patent claim which includes everything in that invention. A claim dominates an invention which contains every element or part mentioned in the claim. Each claim in a patent, then, may be considered as if it were a whole patent. To illustrate, if a claim contains the parts A,B,C,D, and E of an invention which contains parts ABCDE, that claim dominates the invention. If a claim does not read on all the parts ABCDE, it does not dominate the invention because of anticipation by a prior patent, or because it defines and covers only a portion of the invention. A patent containing such a claim, in the absence of any other claim which dominates the invention, can be circumvented.

CIRCUMVENTION

Limiting phrases in the claims of a patent make it liable to modification by others. They have a legitimate right to take advantage of weaknesses in the previous patent and exploit the competing, but noninfringing, device. Thus, getting around the language of a weak patent claim without exposing one's self to infringement is called circumvention. If a circumventing invention is actually better than that dominated by the circumvented patent, the new invention may destroy completely the value of the patent.

Sometimes a patent will illustrate two or more species of an invention, and the claims will read on only one species. In such an event, unless the remaining species are anticipated by other unexpired patents, any part or whole of the remaining species may be exploited to advantage, since they are not covered by the patent claims. Therefore, a prospective patent applicant should evaluate his patent position by asking himself, "Can the claims of my invention get around the language of the broadest of the previous patent claims pertinent to it?" If such an attitude is assumed in devising and developing a product, the invention would be the subject of a valid patent.

5

Patent Application

A patent application is a written document describing an invention in detail by reference to a drawing which usually should accompany the application. The application contains a petition directed to the Commissioner of Patents and requesting that a patent be granted on the invention, giving its title. The patent application should be applied for and signed by the inventor, joint inventors, legal representatives, or assigns of the invention. It also contains a specification describing the invention, a claim or claims, and an oath stating that the applicant is the inventor of the idea or discovery and that he has not seen or has any knowledge of a similar invention patented in the United States or aboard. A complete application for a patent consists of the following, given in the order represented:

1. *Petition*—Usually written on a single page.
2. *Abstract*—Written on the following page, numbered Page 1, appearing just before the description of the invention (specification).
3. *Specification*—As stated above, the specification is the description of the invention, described in concise and readily understandable language.
4. *Claim*—The specification must terminate with one or more claims, which distinctly point out the part or parts of the invention which the inventor claims to be his invention or discovery.
5. *Oath*—An oath written on a separate sheet follows the last claim page.

6. *Signatures*—The inventor must sign the petition, the last page of the specification (which contains the last claim), and the oath, which should be signed before a notary public or other government officer authorized to administer oaths.

PATENT SEARCH

Before an application for a patent is prepared, a patent search for prior art (disclosures) must be made to determine the extent of novelty of the invention and its patentability. In order to accomplish this search, a sketch must be made illustrating the invention, together with a concise description of the invention. This material is usually sent to a patent searcher in Washington, DC or its environs and is accompanied by a fee charged by the searcher. The fee ranges from $6 to $35, depending on the complexity of the invention and on the extent of time necessary to complete the search in the Patent Office for prior patents or publications describing similar inventions or products. The search takes from one to three weeks. If an opinion by the patent searcher is desired, a request should be made to the searcher, and an additional fee must be included. The Patent Office does not make any patent search for the inventor until a formal patent application has been filed in the Patent Office.

When the inventor receives the copies of prior patents or other prior publications from the patent searcher, he must study the disclosures thoroughly to determine if there is any subject matter which anticipates (occurs prior and is similar to) his invention. The inventor must then decide whether his invention possesses sufficient novelty to warrant the preparation of a patent application. If there is not sufficient novelty in the invention, he may decide to take no further action, to save any unnecessary costs. If he thinks there is novelty in the invention, then he should proceed to prepare his patent application.

PARTS OF A PATENT APPLICATION IN DETAIL

In preparing the application, the following order of arrangement should be observed.

Petition

The petition is directed to the Commissioner of Patents, United States Patent Office, Washington, DC 20231. The petition contains the full name of the inventor, his address, and a request for granting of a letters patent for the invention, which is given a suitable title by the inventor. (Sample petitions are given later in this chapter.)

Specification

The specification includes a description of making the invention and the manner of its use. The description must be in full, clear, concise, and exact terms as to enable any person skilled in the art to which the invention belongs, to make and use the same (if permitted). The specification, as well as the petition, claims, and oath, must be typed on legal size paper, preferably 8½ inches by 13 inches. There should be a margin of 1½ inches on the left-hand side and at the top of the pages. Each page should be consecutively numbered, starting with "Page 1," the page numbers being placed in the center of the bottom margins. The material must be typewritten, double-spaced, and each typed line must be numbered from 1 to 32 in the left-hand margin. Such a numbering makes it easier for the applicant and for the examiner when a reference to a part of the specification or claims is made during prosecution of the case.

The specification should begin with the title of the invention typed in CAPITAL letters, placed in the vertical center of the page and in line with the numeral 1 of the left-hand numerals. The title "Abstract of the Disclosure" is centered below the invention title and across from the 2nd or 3rd numeral (also referred to as line). One line below this title starts the content of the abstract (first line indented 6 to 10 spaces).

Abstract—Newly adopted Rule 72(b) requires that each patent application should contain a brief abstract of the invention disclosure in the specification, the abstract to appear immediately after the title of the invention and preceding the descriptive disclosure. It should be a separate paragraph and preferably should not exceed 100 words. The abstract is a condensation of the disclosure, in clear and concise language. The abstract should not contain the relative merits or speculative applications of the invention.

Summary of the Invention—The summary of the invention (with no title given) follows the abstract. The summary includes the nature of the invention and may state one or more prior patents or products relative to the invention, showing their disadvantages compared with the present invention. This could be one short paragraph. Thereafter, in a series of short paragraphs, the summary states the various objects of the invention and explains their implementation, giving valid reasons why these objects (objectives) are advantageous over the existing products. The advantages may be a reduction in the number of parts, use of inexpensive materials, higher efficiency, more flexibility of use, or similar characteristics of the invention.

Reference to Drawings—A brief description of each drawing is made, giving the figure number (starting with figure 1) and preferably, written in one line. For instance, "Figure 1 is a perspective view of my invention," and "Figure 2 is a cross sectional view of the invention," etc.

Body of the Specification—Following the description of the drawings, the main description of the invention begins by reference. For instance, begin with reference to figure 1 and the main structure numbered 1; then describe, consecutively, the numbered breakdown parts of the main structure. After a thorough description by reference to numbered parts of the drawings, a brief description of the operation of the invention is given, indicating or emphasizing certain principal structures which enter into the formation and operation of the invention.

The specification has a dual purpose—one, to describe the invention by reference to the numerical callouts, and two, to form a basis for the scope of the claims embodying the invention. Subject matter not shown in the drawing or described in the specification cannot be claimed in the claim section. Also, the terms used in the claims must have clear antecedent (prior usage) in the specification. For instance, "the housing 24" in the specification cannot be claimed as "the framework" in the claims, unless the specification states, "the housing or framework 24." Also, any functional characteristics of a given part being claimed in a claim should not differ from that described in the specification, unless both functions are given or described in the specification.

Claims

An invention may be claimed by one or more claims, depending on the novelty and complexity of the invention. The claims define the scope of the patentee's monopoly and include the novel features of his invention. The claims must be supported by sufficient disclosure in the specification and in the drawings; otherwise, they may be declared invalid. As many claims as are necessary to fully protect the invention may be included in the application.

Drawing

Whenever the nature of the case admits, the applicant is required to furnish a formal drawing, using India ink. The drawing must be made on a white two-ply or three-ply Bristol board, 8½ by 14 inches. Lines drawn to mark off margins are no longer permitted. The drawing must be made in an area 2 inches below the top margin and ¼ inch from the bottom and side margins, thereby leaving a drawing area of 8 by 11¾ inches.

Drawings are made at a scale large enough to show the fine structures without crowding. Different views, as necessary, including cross-sectional views, should be made, and the parts should be numbered consecutively as the description proceeds. If the same part appears in several views, it must always be called out by the same numeral and name.

Oath

The oath declares that the inventor believes himself to be the first inventor of the particular item or product. It is the last page of the application and is signed by the applicant (inventor) before a notary, who also signs and places his notary public seal on the same page. (Sample oath shown later in this chapter.)

Signature

The petition, the oath, and the specification are signed by the inventor. Frequently, the inventor also signs the drawings at the lower right-hand corner of each sheet, although this is not a Patent Office requirement. The signing of the drawing will be of advantage if the drawing becomes separated from the application because of some corrections to be made on it by the Patent Office draftsman. Again, the inventor's signature on the drawing is arbitrary since other identifications, such as the case serial number and date of filing, are usually placed on the drawing by the Patent Office when it receives the application.

FILING AND FINAL PATENT FEES

The initial filing fee for a patent application is $65. In addition, a fee of $10 for each independent claim in excess of one, and $2 for each dependent claim in excess of 10 claims must be paid at the time of filing the application. Any errors in payment can be rectified by

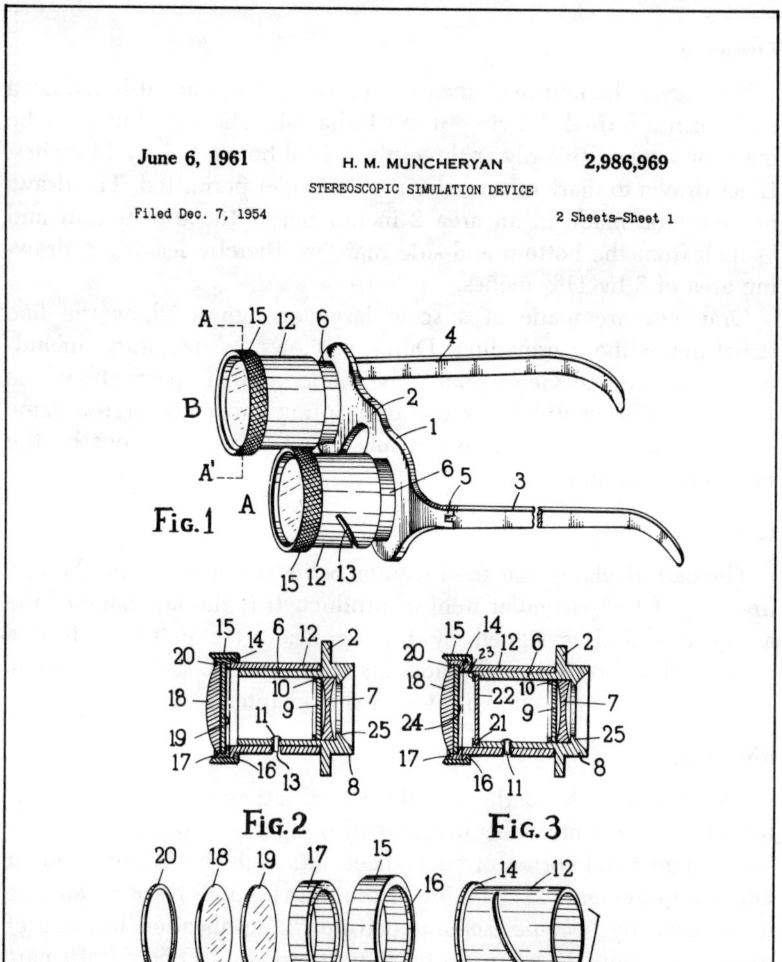

Diagram showing a perspective view, a cross-sectional view, and an exploded view. (The last view is not required in all drawings.)

May 14, 1968 H. M. MUNCHERYAN 3,383,491

LASER WELDING MACHINE

Filed May 5, 1964 4 Sheets-Sheet 4

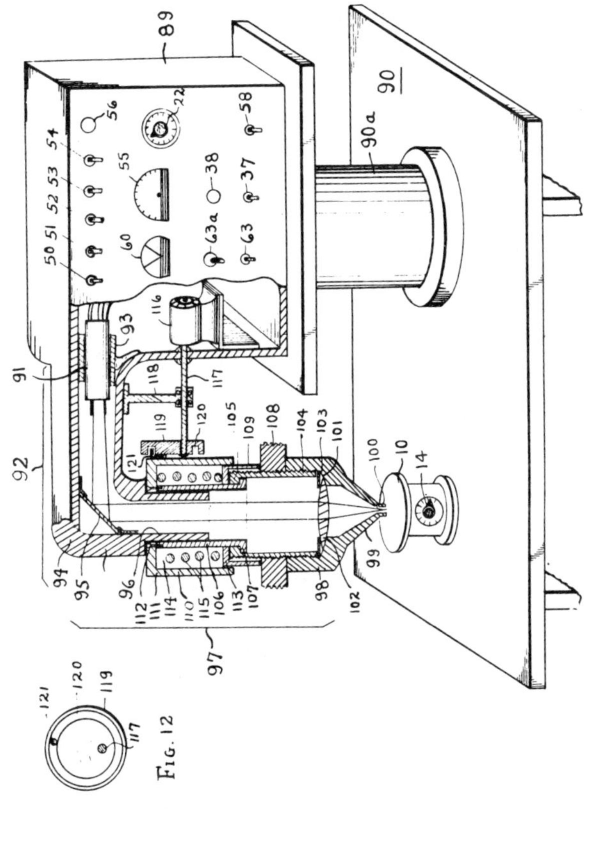

Perspective view of a laser welding system showing the instrument panel and the laser head partially cut out.

May 7, 1968 H. M. MUNCHERYAN 3,382,343

LASER WELDING MACHINE

Filed July 23, 1964 5 Sheets-Sheet 5

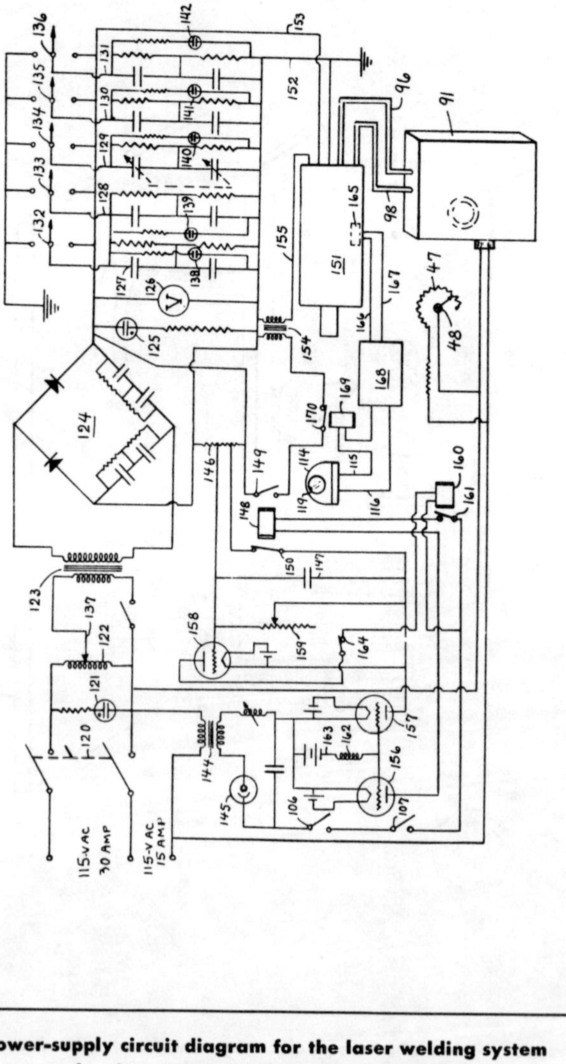

Power-supply circuit diagram for the laser welding system showing conventional electronic symbols.

the Patent Office. That is, if an excess payment is made, the Office returns the excess; if insufficient payment is made, the Patent Office notifies the applicant to make the additional payment before the application is officially admitted.

A final fee of $100, plus $10 for each page of the specification printed and $2 for each sheet of drawing, must be made upon allowance of the application. To avoid lapsing of the patent application, the final payment must be made within three months after notification of allowance is made.

INVENTOR'S MONOPOLY IN A PATENT

All that is disclosed in the specification and drawings of a patent becomes public knowledge. Only that portion of this knowledge which remains as monopoly of the inventor is defined by the claims. Strength of a patent is determined by the scope of the claims as supported by the inventor's disclosure in drawings and the specification.

SUMMARY OF INSTRUCTIONS

The inventor must first write the Petition, in accordance with the proper form (given later). If the application is by joint inventors, each of their names appear on the Petition, and the proper form for joint inventors is used. Both inventors sign the Petition, giving full names (first name, middle initial, and last name).

The Abstract appears on the second sheet of the application, marked as Page 1. In preparing the rough draft of the application, the writing of the abstract may be delayed until the entire application, including the claims, is completed. This facilitates the writing of the Abstract, because some lines from the specification and claims may be lifted and incorporated into the Abstract, making it concise and to the point.

The next area is the Summary of the invention. It does not contain any reference to the parts of the drawings by numeral. Statements in the summary are generalized and brief.

Following the Summary, references to the drawings are made. Each figure title is referenced consecutively and ends in a period, except the next to last one. This particular figure title terminates in a comma followed by "and."

The body of the Specification references and defines every significant part in each drawing, by name and function. This section may begin, "Referring now to the drawing, reference numeral 1 designates . . . having xxx (part) 2 connected to xxx (part) 3 adjacent thereto . . . ." This description states how the product is built, giving the arrangement of parts in the construction and describing the interrelation of the parts. This forms the basis of the claims to be subsequently drafted. A brief discussion of the operation of the device is given near the end of this section. It is to be remembered that the drawings and parts are to be numbered in the same progression as the written description. To ensure this, it is best to make a copy of the drawings and place the numerals on the copy as the description is written. This will avoid making any changes or corrections on the formal drawings on the Bristol board.

The claims of the invention are drafted after the callout numerals and descriptions have been completed. Since each claim is considered as an individual patent on the part or parts it recites, preparation of the claims should be done with great care. (A detailed instruction, giving examples of various types of claims and their importance, is presented in Chapter 6.) In drafting the claims, it is important to use words, names, and clauses that find antecedent in the specification and/or in the drawings.

After the claims have been double checked, an oath is prepared in accordance with the form given later in this chapter.

Various forms for Petition and Oath follow.

PETITION FOR PATENT BY A SOLE INVENTOR

To the Commissioner of Patents:

Your petitioner, ..., a citizen of the United States and a resident of . . . ., State of . . . . (or, subject, etc.) whose post-office address is . . . ., prays that letters patent may be granted to him for the improvement in . . . ., set forth in the following specification.

(The specification and oath follow the petition.)

PETITION FOR PATENT, BY A SOLE INVENTOR, FOR HIMSELF AND ASSIGNEE

To the Commissioner of Patents:

Your petitioner, ..., a citizen of the United States and a resident of . . . ., State of . . . . (or subject, etc.), whose post-office address is ...., prays that letters patent may be granted to himself and ...., a citizen of the United States and a resident of ...., State of ....,* whose post-office address is ...., as his assignee, for the improvement in ...., set forth in the following specification.

PETITION FOR PATENT BY JOINT INVENTORS

To the Commissioner of Patents:

Your petitioners, ... and ...., citizens of the United States and residents, respectively, of ...., of State of, ...,* and of ...., State of ....* (or subjects, etc.), whose post-office addresses are, respectively, ... and ...., pray that letters patent may be granted to them, as joint inventors, for the improvement in ...., set forth in the following specification.

PETITION FOR PATENT BY AN ADMINISTRATOR

To the Commissioner of Patents:

Your petitioner, A............. B............., a citizen of the United States and a resident of ...., State of ....* (or subject, etc.), whose post-office address is ...., administrator of the estate of C............. D............., late a citizen of the United States and resident of ...., State of ....,* deceased (as by reference to the duly certified copy of letters of administration, hereto annexed, will more fully appear), prays that letters patent may be granted to him for the invention of the said (C............. D.............) for an improvement in ...., set forth in the following specification.

OATH TO ACCOMPANY APPLICATION FOR PATENT

(The petition and specification precede the oath.)

(1) ...., the above-named petitioner ...., being sworn (or affirmed), depose ... and say ... that ....

---

*Include the ZIP code after the name of the state.

citizen .... of the United States (2) and resident ..... of (3).............., that .... verily believe .... (4).............. to be the original, first, and (5) .... inventor of the improvement in (6).............. described and claimed in the foregoing specification; that (7).............. do .... not know and do .... not believe that the same was ever known or used before (8).............. invention thereof, or patented or described in any printed publication in any country before (8).............. invention thereof, or more than one year prior to this application, or in public use or on sale in the United States more than one year prior to this application; that said invention has not been patented in any country foreign to the United States on an application filed by (9) .... or (8).............. legal representatives or assigns more than twelve months prior to this application; and that no application for patent on said invention has been filed by (9) .... or (8).............. representatives or assigns in any country foreign to the United States, except as follows: (10).....................,

Inventor's full)...........................
name. (11))...........................
(Signature)

..............................)
) ss:
..............................)

Sworn to and subscribed before me this ..... day of ...., 19......,

(Seal)

....
(Signature of notary or officer)

(12) ....
(Official character)

OATH NOT ACCOMPANYING APPLICATION

..............................)
) ss:
..............................)

.............................., being sworn (or affirmed), deposes and says that he is a citizen of the United States of America and resident ...., that on ...., 19......, he filed application for patent Serial No........ in the United States

Patent Office, that he verily believes himself to be the original, first and sole inventor of the improvement in .... described and claimed in the specification of said application for patent; that he does not know and does not believe that the same was ever known or used before his invention thereof, or patented or described in any printed publication in any country before his invention thereof, or more than one year prior to the date of said application, or in public use or on sale in the United States more than one year prior to the date of said application; that said invention has not been patented or made the subject of an inventor's certificate issued before the date of said application in any country foreign to the United States on an application filed by him or his legal representatives or assigns more than twelve months prior to the date of said application; and that no application for patent or inventor's certificate on said invention has been filed by him or his representatives or assigns in any country foreign to the United States, except as follows: .....

Inventor's full name: ....

....

Sworn to and subscribed before me this ...... day of ...., 19......

____

(Signature of notary or officer)

____

(Official character)

KEY TO NUMERALS AND LETTERS IN PRECEDING SAMPLE FORMS

Numerals

(1)—Name of inventor, or names of joint inventors.

(2)—If applicant is an alien, state foreign country of which he is a citizen.

(3)—Give city and state, or if a foreign resident, city and country, of residence.

(4)—"Himself" for sole inventor, and "themselves" for joint inventors.

(5)—"Sole" in the case of sole inventor; "joint" in the case of joint inventors.

(6)—Title of the invention.

(7)—"He" for sole inventor; "they" for joint inventors.

(8)—"His" for sole inventor; "their" for joint inventors.

(9)—"Him" for sole inventor; "them" for joint inventors.
(10)—If no application has been filed in a foreign country, strike out the word "except as follows:"

Letters

A—First name(s) of executor(s), and middle initial(s).
B—Last name(s) of executor(s).*
C—First name of the inventor, and middle initial, if any.
D—Last name of inventor.

NOTE: For a complete patent application, please see the example presented in the Appendix.

---

*Executor or administrator, used synonymously.

6

Patent Claim Preparation

GENERAL

A patent claim defines the scope of the patentee's monopoly, pointing out definitely what he claims as his invention. By the issuance of a patent, the specification and the drawing become public knowledge, and all the inventor reserves for himself are certain rights limited by the strength of his claims. A claim dominates everything that it includes. If more than one claim occurs in a patent, each of them may be considered to constitute a separate scope of the invention disclosed. Therefore, a patent is only as good as the claims which define it.

PREPARATION OF CLAIMS

To write a good claim, one requires imagination, foresight, and a high degree of knowledge of the product being patented. He must also have the ability to present abstract concepts of the invention in precise language. To achieve this professionalism takes practice; write the claims over and over, if necessary, until a concise and crisp operation description of the part, or parts, being claimed results. The main purpose of this section is to point out the importance of understanding the scope of the claim(s), so that the information may be properly applied to fully cover the claims of the invention. It is also necessary for the claim writer to be able to determine whether his project is infringing on prior patents; he must know what steps should be taken to avoid infringement or to circumvent the prior patents. This is very important to the inventor; time and money are often spent to develop a product only to find out that it has been anticipated (covered) by prior patents. For this reason, the value of a patent search at the onset of a development cannot be overemphasized.

The patent search has several benefits. The search will disclose what prior information exists; whether the invention has been anticipated; and whether the prior art is so old that the patentability of the invention is doubtful. The prior art uncovered sometimes teaches the inventor a better way to make his invention or to improve certain parts of the already developed invention. From the search, the patent writer learns (if he does not know already) the terminolgy used in a similar line of products and the broad terms used to cover the invention.

Prudent innovators and developers of products are learning more and more by making inquiries on their own as to whether a certain part they intend to use in their development is "safe." This should be practiced by all innovators, design and development engineers, and researchers. It goes without saying that this practice is routine in many prominent research laboratories. Rather than using "brute force" on the mental ingenuity to make a start on a new problem, it will be easier to uncover, through patents, how others have achieved the same end, difficulties they have encountered, and what means they have used or will be necessary to overcome these difficulties. The specification will describe the methods used, and the claims will specify the means monopolized by the patentee to accomplish the methods.

TYPES OF CLAIMS

As a general rule, claims may be classified into generic claims and specific claims. These, in turn, can be divided into combination claims and subcombination claims. A generic claim has broad construction and includes every limitation of the invention, element by element. It is usually shorter than a specific claim and broader in scope. For example, the words "television tubes" refer to all types of television tubes, regardless of size and color. But, "24-inch television picture tubes" refer to a smaller number and specific type of television tubes. If we say "25-inch color-television picture tubes," a still more specific and smaller number of television tubes are included, because it does not include the black-and-white television picture tubes. Thus, every word or phrase added to define a claim narrows it down to a more specific structure. Therefore, if a discovery or improvement has been made in a 24-inch color-television picture tube, it will be preferable to refer in the claim simply to a "television tube" to cover a broader ground in the claim, if no prior art exists.

The more crowded the prior art is, the more limiting phrases will be required in a claim to narrow it down and to avoid the prior art. On the other hand, when a claim is too narrow it will thereby be worthless.

There are two principles which should not be overlooked in studying a patent. The first is that anything disclosed in the drawing or described in the specification, but *not* claimed in the patent, is public property, to which the patentee has no exclusive right. Second if the patentee, during the course of prosecuting his invention in the Patent Office, has made a claim and thereafter abandoned it by cancellation, the issue of a patent without claims of similar scope specifically abandons all rights to the cancelled matter. This can be determined by anyone purchasing the copies of the patentee's application and prosecution file from the Patent Office and studying the official actions transacted. This is always done for adjudicating (to settle or determine by the decision of a court) an issued patent. Also, to circumvent a patent, the application file of the particular patent should be procured and thoroughly studied. Very often a company would be paying license fees to a patentee whose patent, when adjudicated, may actually be held invalid on the above-explained grounds.

Some of the principal features of a claim may be explained by a hypothetical invention: a table with four legs and a central drawer for holding stationery or other articles.

Broad Claim

"A table comprising a top section having means for holding stationery and second means for supporting said top section a suitable distance from the floor."

This claim will dominate any type of table having a top section with means for holding stationery and second means for supporting the top section a suitable distance from the floor. For instance, it will cover the following features in a table:

1. A table having center-guided drawer or drawers built under the top section.
2. A table having center-guided drawers built under and on top of the top section.

3. A table having a drawer or drawers on one or both sides of the table underneath the top section, such as a desk would have.
4. Instead of four legs, the table can be made with three sides with leg space between the left and right sides.
5. The claim covers a typewriter table with four legs and a central drawer, or a table with one leg.
6. A boudoir table is covered by this claim.
7. Any table with or without drawers.

It will be seen that if such a claim is allowed, all tables having one or more drawers and means for supporting the table top a suitable distance from the floor will be covered by the patent having this claim. Consequently, it will be difficult to build a table, if not impossible, without the features explained above; this claim, therefore, cannot be circumvented.

Narrow Claim

"An oblong table having a top section and four legs for supporting said top section a suitable distance from the floor, said top section having four center-guided drawers incorporated on the upper and back side thereof, and said center-guided drawers being arranged in a straight line so as to form a linear shelf space thereupon."

This claim is so narrow that it can be easily circumvented by making a table with the following features:

1. A table having three legs instead of four legs.
2. A table having three sides with curved shelf or drawer construction.
3. The drawers may be incorporated under the table top, thus increasing the work space on top of the table.
4. In a three-sided table, the drawers can be made under the table top and on one side only.
5. In a three-sided table, the drawers can be incorporated on two sides of the table with an included space for legs of the user.
6. In a three-sided table, side drawers and a central drawer can be made, with leg space between the side drawers.
7. In a three-sided table, a large number of drawers can be made and installed at the top and back side of the table top; or instead of incorporating drawers, a large number of shelves or one shelf for books can be incorporated to circumvent the narrow claim.

It is thus seen that a narrow claim of the character presented above is a worthless claim, because it gives monopoly on one type of table and restricts the market to the particular type of table. In the broad claim, a table top held up by wires or ropes attached to the ceiling of a room will be dominated.

Broad Claim

"An electric motor comprising a stator and a rotor, means secured to said stator for clutching said rotor, and means secured to said rotor for cooling said motor during operation."

Such a claim will dominate all types of electric motors, all types of clutches used with a motor, and any method of cooling.

Narrow Claim

"A direct-current motor, comprising a stator, a rotor included therein, and a rotatable shaft with a circular plate disposed at one end thereof being axially connected to said rotor, a second plate secured to said stator in juxtaposition to said first plate for frictional engagement therewith, and a fan blade secured to the opposite end of said shaft for blowing air on said motor to dissipate heat therefrom."

This narrow claim is of very limited scope and of doubtful patent value, because:

1. Instead of a dc motor, one may use an ac motor, induction motor, or a two- or three-phase motor to circumvent the claim.
2. Instead of a circular plate for clutching, one may use a split-ring or ferrule-type clutch mechanism.
3. Using the mechanism in (2) will eliminate the use of the first and second plates. This would be considered an additional improvement.
4. Instead of a fan blade, the motor may be provided with a wheel to blow air, a tube with a shower head to blow air or other fluid, or the motor may be immersed in oil or in some type of fluorochemical circulated by convection current or by means of a pump.
5. The claim is of doubtful patentability because other devices also use plates as clutching means, fan blades for blowing air, or direct current energizing a rotor with a shaft, etc.

Why then would the broad claim be patentable? Doesn't that claim cover all these parts mentioned? Yes, it covers all these parts, but does not name them. The word "means" could be any device to produce the function to be served, provided that no analogous structure has been disclosed in a prior patent or publication older than one year prior to the filing of the application.

Why does one year have anything to do with the filing of a patent application? The patent rule states that any invention filed in the Patent Office within a year after its publication will not constitute a bar to patentability of the invention, provided that all other rules of patentability have been met. The Oath shown in Chapter 5 will further confirm this patentability rule.

DRAFTING PATENT CLAIMS

As pointed out previously, a patent claim distinctly defines the subject matter which the applicant regards as his invention. It is written out as a single paragraph containing a number of clauses and phrases with no period until the last clause or phrase is concluded. The following generalized rules may be stated for drafting claims:

1. A claim must recite an operative structure. For example, "a metal surface" with no function cannot be patented; but, "a reflecting metal surface" in a device can be patented if no prior art in an analogous device exists. The surface has function by reflecting light.
2. If more than one claim is recited in an application, each claim should be written patentably different from the other. What is "patentably different?" For example, if a device to be patented has the parts A, B, and C, and it becomes necessary to write three claims, then one claim should contain part A, the second one part B, and the third one part C. Or, if all three can be included in one claim, then that claim should contain parts A, B, and C.
3. A claim must define exactly the structural features and the relationships of the structures. The relationship must be such that the features operate cooperatively. A claim for "a writing device, comprising a pencil having a rubber eraser at one end," is rejectable. There is no cooperative function between the pencil and the rubber eraser; each can be used independently of the other. A claim on "a cutting device, comprising a handle having a blade secured to said handle," is patentable in the absence of prior art. The blade cannot be used without

the handle, and the handle cannot be used for cutting without the blade. The two cooperate operatively with each other.

4. A claim must set forth positively the principal elements contained therein. Rather than stating "a chair having a leg supporting it a suitable distance from the floor," one would state "a chair comprising a seat and a leg supporting said seat a suitable distance from the floor." Both the "seat" and the "leg" are positively recited.
5. A claim must recite the parts around the hole rather than the hole itself directly. State the claim as "a disc having an aperture in its center," and not as "a hole contained in the center of a disc," because a hole is not an element; it is nothing, and a claim cannot be drawn on nothing.
6. A claim must not contain a mere aggregation of parts, but it should indicate how each part cooperatively brings about the intended result. For example, "a lamp comprising a base, a goose-neck, an electric bulb, and a five-foot electric cord" is not patentable because it is an aggregation of parts showing no cooperative functions between the parts. "A lamp comprising a base, a goose-neck secured at one end to said base, an electric lamp secured to the other end of said goose-neck, and an electric cord carrying current therethrough being connected to said bulb through said base and said goose-neck to energize said bulb" is a claim subject to patentability if no prior disclosure exists.
7. A claim should set forth the combination of means by which the result is obtained rather than the mere function or principle. "A lamp to be used on a desk to light the top of the desk" is a functional claim and cannot be patented, because it does not recite the parts which produce the particular function. Refer to the claim in (6) for correct answer.
8. There should be no duplication between any two claims without reciting or appending a new part in the second claim. For example, in the claim presented in (4), a second claim may be written exactly the same way by inserting at the end of the clause, "said seat having a back projecting outwardly therefrom." This claim differs from the other because the chair has an additional part, which is the "back" projecting.
9. A claim should not rely on reference characters (numerals) for describing the different parts or structures of the invention. Early history of patents shows that this was common practice,

but that rule has now been changed. No numerals are used in a claim to define the parts.

10. A claim should not state the elements alternatively; that is, ". . . said laser-generating element is either a ruby or a gallium arsenide rod." However, it could state that, ". . . said laser-generating element consisting of one of the general classes of semiconductors characterized by gallium arsenide, indium arsenide, and silicon carbide." Always be positive!
11. A claim must not read on prior patents, whether they belong to the applicant or to other inventors.
12. When it is difficult to state the element (part) positively, certain negative words such as "nonferrous," "noncircular," "nonrectilinear," etc., can be used. For example, if an alloy does not have any iron in it but contains aluminum, copper, and silicon, then it can be referred to as "nonferrous" alloy. As another example, when a part is not circular but has any other shape, such as partially rectangular, oval, irregular, or anything between that cannot be defined, the term "noncircular" may be used.
13. A claim should not include something which is not necessary in bringing about the desired result, as in the statement, ". . . said hammer (of the gun) triggering the explosive, <u>whereupon the smoke comes off the barrel.</u>" The underlined clause is not necessary to bring about the operation of the hammer or formation of explosion.

Specific and Generic Claims

The following types of claims are identified in an application and referred to by the Examiner during prosecution of the application.

Specific Claim—A specific claim is a narrow or restricted claim and covers the part or structure directly and in detail. A specific claim may be worded, ". . . said housing having an incandescent lamp energized by a direct current source," or, ". . . said housing having a neon lamp energized by an alternating current." Each of these claims, or parts of a claim, describe explicitly the types of lamps and the types of currents energizing each type of lamp. Therefore, these two claims cover only these two types of lamps and the two types of currents used. Any other type of lamp or current (for instance, the pulsing type) is not covered by these claims. Accordingly, a generic claim must be written in order to cover all types of lamps and all types of currents.

Generic Claim—A generic claim broadly covers the principle of an invention, including in its scope a number of specific claims. An example of this type may state, ". . . said housing includes a radiant means energized by a current." This type of claim includes all types of radiation-producing devices (lamps) and all types of currents, whether the current is direct, alternating, pulsing, or oscillating.

Here, we must also introduce a new term, "species." Species means a group of things having one or more characteristics in common. In an application, when a drawing represents the principal device and a second drawing represents a modified form of the principal device, then the application is said to have two species, each being the modification of the other.

DIFFERENT SPECIES IN ONE APPLICATION

More than one species of an invention, not to exceed five species, may be claimed in a single application. In such a case, the application must contain at least a single generic claim which covers all the claimed species. That is, a broad claim should be written in such a way that it will cover every part (limitation) of the given species. For instance, in the foregoing examples of specific and generic claims, if two modifications of the same invention were shown in the drawings, then a generic claim must be prepared to cover both of the modifications (species), as demonstrated above. All the other claims for each species are written in a form dependent on the generic claim. Several examples will illustrate this point.

An example from Patent No. 2,192,569, for a magnetic cigarette and support therefor, is illustrated in this text. In this invention, there are two species of a cigarette. Each serves the same purpose. The cigarette has at one end a magnetizable material. When the burning cigarette is placed in an ashtray having a magnet, the cigarette does not fall over; the magnetic end holds it firmly against the ashtray. It will be seen that in one species, the magnetic material is in the form of a wire, and in the other the magnetic material is in the form of a ring. Both materials can be steel or an alloy of iron, so that it becomes magnetized when brought near a permanent magnet, such as the U-shaped part secured to one side of the ashtray. Unless a generic claim is written to cover both of these species, they cannot be patented in one application; two separate applications must be filed. However, a generic claim can be written to cover both of these species, provided no prior art exists.

Generic Claim—"The combination of an ashtray and a cigarette, comprising an ashtray having a recessed portion with a magnet secured at one side thereof, and a cigarette provided at one end with a magnetizable means adapted to be accommodated on said magnet and being stabilized thereby when placed to rest thereon during smoking." This claim may be marked as Claim 1.

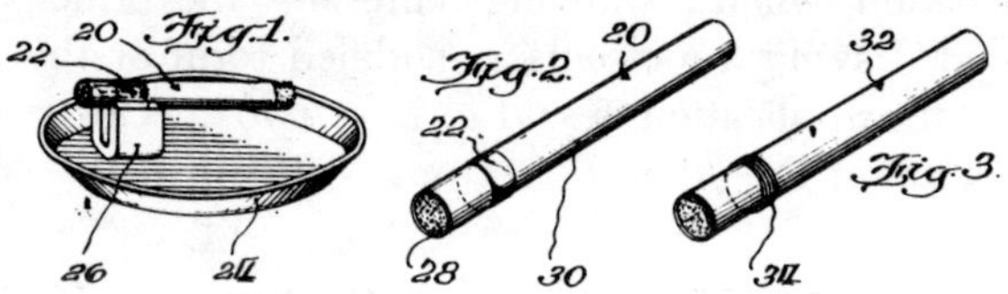

Magnetic cigarette and support therefor: Patent No. 2,192,569

The generic claim covers any kind and shape of ashtray having a magnet (any shape) adapted to hold a lit cigarette, which is provided at one of its ends (preferably near the unlit end) with a magnetizable material. This material can be a flat piece of magnetic steel, a magnetic wire, a magnetic ring, or the like. In the drawing, two types of magnetizable means are given: one is a ring, and the other a coiled wire. However, a "magnetizable means" covers any convenient shape of material that can be held by the magnet 26 in the drawing. In Fig. 2, the cigarette 30 is provided with a magnetic ring or annulus, and in Fig. 3 the cigarette 32 is provided with a coil of wire made of some magnetic material, such as steel. Both the ring and the wire are magnetizable when brought in contact with the magnet secured to the ashtray. If the generic claim is allowed by the Patent Examiner, then species claims must be written to cover each of the two species. The species claims will be:

Species Claim No. 2—The combination of an ashtray and a cigarette as defined in Claim 1, wherein said magnetizable means provided at one end of the cigarette is an annulus made of steel.

Species Claim No. 3—The combination of an ashtray and a cigarette as defined in Claim 1, wherein said magnetizable means provided at one end of the cigarette is a coil of wire made of steel.

A patent allowed with these three claims can be circumvented, because in this patent the permanent magnet is secured to the ashtray, and the magnetizable material is secured to the cigarette. Suppose the annulus and the coil of wire are made of a permanent magnet material, and the magnetizable material is provided on the ashtray; then the principle of magnetic holding and stabilizing will be changed. Thus, a second applicant may get a patent on his invention, although the examiner may bring objection that the first patent is teaching the principle of magnetic holding of a cigarette. This objection may be circumvented by stating in the amendment (remarks section) that the cigarette with the permanent magnet can be placed on any steel structure, such as a steel bench, table, beam, or any other part, to stabilize the cigarette; this is an added advantage over the previous patent (No. 2,192,569). Therefore, it is said that the first patent does not have a strong patent posture. Accordingly, the generic Claim 1 should have been written as follows:

Generic Claim No. 1—The combination of an ashtray and a cigarette holder, comprising an ashtray having a recessed portion provided with a magnetic holder means secured at one side thereof, and a cigarette holder of magnetic means secured at one end of a cigarette and adapted to be accommodated on said magnetic holder means of said ashtray and to be stabilized thereby when placed to rest thereon during the interval when said cigarette is not being smoked.

This generic claim, if allowed when no prior art exists, covers a cigarette with a magnetized or nonmagnetized annulus or coil of wire, and an ashtray having a magnetized cigarette holder or nonmagnetized cigarette holder of magnetizable material. Thus, either the annulus (or wire) or the holder in the ashtray can be a magnet. Such a claim is not subject of circumvention. Then, it will require four species claims instead of two because of the variations in the location of the magnet and the variations of the forms of magnetic means on the cigarette. The reader should try to write these four species claims following the explanation given above. In an application containing a generic claim and species claims, the generic claim is also known as *Independent Claim,* and the species claim is known as *Dependent Claim,* since the latter is written in a form dependent on the generic claim.

DIFFERENT SPECIES CLAIMS IN ONE APPLICATION

A patent application embodies only one invention, with modifications up to five species. If a patent application containing several species does not have a generic claim, the Patent Examiner will declare and require the application to be restricted to one species, elected by the inventor. This means that a new application will have to be filed for each additional species; this is costly and time-consuming. Therefore, the importance of a generic claim in a multiple-species application cannot be overemphasized. Each of the new (divided application) applications will bear the filing date of the original application; this is an advantage. All divisional applications will consist of the drawing, specification, and claims of only those parts that are contained in the divisional application. The divisional application is handled in the same manner as an original application for a single invention.

Let us assume that no generic claim was found in the application for the ashtray and the cigarette holder. Then, the original application may pursue the invention shown, for instance, in Figs. 1 and 2. The second divisional (continuing) application may discuss and claim the invention shown in Figs. 1 and 3. Both of these applications will have the original filing date. The first application will state in its preamble, "This application was divided into a divisional application, now Serial No. . . . (new serial number)." The second application will contain in its preamble, "This application is a continuation of the application filed on . . . (date of filing) and Serial No. . . . (give serial number)." This way, the two applications will tie up with each other and be properly identified. Claiming and prosecution of both of the applications will be the same as when they were originally separate applications for separate inventions.

A complete application for an automatically heated ice-cream scoop with stand is shown in the Appendix of this book. It will be seen that prior arts covering parts of this invention are cited at the end of the patent. Depending upon how broadly or specifically the prior art covers any given invention, the claims, including the generic claim, have to be narrowed by use of defining words or phrases (as discussed earlier). This is done so that the invention will cover specific details of construction and will circumvent prior art. A patent of this type will cover only what is claimed in the claims, since the art has been practiced quite widely. The principal distinction of this patent from the prior patents lies in the fact that no prior patent had shown any stand for the ice-cream scoop. This particular patent also has an advantage over prior ice-cream scoop patents in that it has no dangling electric wires or dangerous heat sinks.

7

Prosecution of Patent Applications

FILING THE APPLICATION

After a patent search has been made and no prior art to anticipate on the invention has been found, a patent application consisting of a petition, oath, abstract, specification, claims, and drawings, accompanied by a filing fee, is filed in the Patent Office. Within a week or ten days after filing, the Patent Office sends the inventor a filing receipt showing the serial number, filing date, and the division or group to which the case is initially assigned. Thereafter, the inventor, if practicing the invention, may mark his articles 'Patent Applied For," but in no case "Patent Pending." A patent is said to be pending only after at least one official action has been received.

In recent years, the time between the filing date and the first action has been from five months to several years, depending on how busy the assigned Patent Office division is. The application is assigned to an Examiner, who examines and studies the application to understand fully the construction and operation of the invention. He then makes a search of all pertinent art to uncover patents, articles in publications, or related sections in a textbook, and makes a comparative study of the invention and the prior art references. When his study is completed, he prepares an official letter to the inventor in which he discusses prior art with reference to the invention. This is known as the first official action.

THE FIRST OFFICIAL ACTION

The first official action contains references to the prior patents listed as either "Art Relied Upon" or "References of Record." This may sometimes be followed by another list with the title "Other

Pertinent Art." Then the examiner's letter follows with comments and discussions pertinent to the drawing, specification, or, generally, to the claims. If the examiner feels that the drawing or the specification is not complete enough to enable one to understand the invention, he may reject the application on grounds of insufficient disclosure and may require amendments to be made on the defective sections. Even if no comments are made on these parts, the examiner, in practically all cases, may present objections and rejections to the claims on grounds of prior art, aggregation, indefiniteness, and other shortcomings. He may give one or more of these reasons for his rejection. The inventor has three months from the date of the official letter to respond to this action. If the inventor fails to respond within this period, the application becomes abandoned.

STUDY OF OFFICIAL LETTER BY THE INVENTOR

By studying the references cited by the examiner, the inventor can form an opinion as to how novel his invention is and to what extent the application can be amended to warrant further action on the application. The inventor usually furnishes the examiner with a written statement setting forth the differences, if any, between his invention and the references cited. In his statements he emphasizes the advantages of his invention over those shown in the prior art. These statements must be convincing to the examiner.

Changes may be made in the specification or drawing if no new inventive matter is introduced in so doing. The applicant may amend or cancel his claims and add new claims to broaden or limit them. The applicant must fully respond to each objection in the official letter; otherwise, the application may be held abandoned. If the examiner has combined a number of references to anticipate a claim, the structural differences of the invention should be considered, as well as their differences of function and result. Even when the structural differences are lacking, but the combination of the several references in an inventive manner produces a new function or result, the claim under question should be allowed. Therefore, the applicant should present his argument from this standpoint.

APPLICANT'S AMENDMENT

The inventor must set forth in detail what he believes to be the differences between his invention and those shown in the patents furnished to him, by the Patent Office, for study. Thus the inventor, equipped with detailed information gained from the references cited, will advance a counter argument if he believes that the prior art does not constitute a proper anticipation. He may advance a number of reasons to save the claim or claims which he thinks are allowable. He may give one or more of the following types of arguments:

1. The new improvement (his invention) was not secured until made possible by the applicant.
2. The new invention answers all reasonable requirements over the prior art.
3. The prior patent or patents were issued less than one year before the application for his invention was filed, and that the prior patents disclose but do not claim the invention in question.
4. Examiner's objection is not clear, and further explanation is necessary.
5. The new invention, even though structurally the same as in the prior patents, is intended for a nonanalogous application.

If the applicant agrees with the examiner's objections, he may modify the claim or claims to limit the scope of the invention. He may even cancel some claims, but in doing so he states that the claims are merely being withdrawn in order to present them in a revised and better form. He may also substitute other claims for those cancelled.

No new inventive matter may be entered, however, in any amendment in a given application, except by a divisional application. Several official actions and amendments may be necessary before a patent is granted. A patent is granted on first official action only in rare cases.

REVIVAL OF ABANDONED APPLICATION

When the applicant fails to take the required action during the statutory period (i.e., during the three months allowed for him to take action), the application becomes abandoned, as stated earlier. However, an application thus abandoned may be revived as a pending application if it can be shown that the delay was unavoidable. Acceptable reasons for the delay may be that the applicant was hospitalized, absent from home due to important business, misplaced the application, or for some other personal reason of unavoidable character. It might be that the applicant could not collect all the references cited or disclosed on time to prepare the amendment. Also, the applicant may have mailed the amendment in time to reach the Patent Office, but the amendment may have been delayed in the mail. All delays must be determined by the Patent Commissioner to have been unavoidable.

When it is decided that the application should be revived, then the applicant must make a petition to the commissioner, stating all the facts that caused the delay in responding to the examiner's official action. The petition to the commissioner is simple in its formality. Starting with the first page of the petition, the following form is used:

PETITION TO REVIVE AN ABANDONED PATENT APPLICATION

In The United States Patent Office

John E. Doe	)	Date ____________
Serial No. 222,444	)	
Filed, January 5, 1972	)	Patent Group 213
For: Microwave Communi-	)	
cation System	)	

To the Commissioner of Patents:

In response to the Office Letter dated March 14, 1973, wherein the Examiner holds the Applicant's patent application as abandoned because of the expiration of the statutory period, the Applicant submits that the delay was unavoidable for the following reasons:

(A reasonable cause or causes for delay must be given here.)

A favorable reconsideration of the case is respectfully requested.

City, State ____________
Applicant's Full Name

Date
____

A petition to revive an abandoned application should be accompanied by a fee of $15 payable to the Commissioner of Patents. The fee must be in the form of a postal money order or a bank certified check. The Patent Office may also accept a personal check, but this might delay the case until the check is cleared.

Verified presentation of the causes of the delay usually is sufficient to revive the abandoned application. In infrequent cases where the commissioner has denied the petition, the easiest and least expensive alternative would be to reapply for the invention by using the same drawings, specification, and the claims. The new application should include all the parts (petition, specification, claims, and oath) in the original application, with the exception of the drawings. A separate letter, included in the application, should be directed to the commissioner requesting him to have the old or original drawings become a part of the new application. Naturally, the required fee of $65 must accompany the application for filing.

The new application may include exactly the same specification and claims as it did at the time of abandonment, or the applicant may make some changes, if he so desires, in the specification or claims, without departing from the scope of the invention shown in the drawings.

8

Divisional Applications

NATURE OF DIVISIONAL APPLICATIONS

A divisional application is a continuation of the parent or original application. It may result from several causes during the prosecution of the parent application.

1. A parent application may contain as many as five species of the same invention. In order to cover all these species in one application, one or more generic (broad) claims are drawn, each one of which reads on all the five species and contains the limitations of all five species. Sometimes, the applicant finds it difficult to have a generic claim allowed by the examiner. In such an event, those species that cannot be covered in a generic claim are required to be presented in a divisional application. If the divisional application is filed before a patent is granted on the parent application, the filing date of the parent application can be conserved in the divisional application; this is a distinct advantage to the inventor. The divisional application includes, as usual, the petition, oath, the parent specification (with inapplicable parts preferably removed), the drawings in the parent application (or, sometimes new drawings, as the case may be), and the filing fee.

2. Sometimes an inventor may include more than five species in a parent application in order to take advantage of the filing date of the parent application by conserving that date in the divisional application. In such an event, and provided that the five species have been allowed by the examiner, those exceeding five species are filed in a divisional application bearing the original date of the parent application.

3. In a parent application, some of the claims may cover one invention and others another invention, in accordance with the draw-

ings accompanying the application. In such a case, the examiner requires that the applicant elect one species in the drawing and present all claims reading on that species. The remaining claims are required to be withdrawn from the case. If the applicant or his assigns cannot present an acceptable counter argument, he elects one species, and files the other species in a divisional application or applications, as the case may be. All divisional applications are treated the same as the parent application in the Patent Office.

4. In another case in which two or more distinct species of an invention have been filed in a parent application, and the applicant has been required by the examiner to restrict his claims to an elected species which later becomes abandoned because of final rejection, the other species may be filed in a divisional application before the parent case becomes abandoned.

A continuation application which has not resulted from restriction of the parent application is not a divisional application and will not necessarily bear the filing date of the application first filed. Such an application usually becomes a fortifying patent.

REQUIREMENT FOR RESTRICTION TO FIVE SPECIES

When a requirement is imposed upon an applicant who is claiming more than five species, the applicant must cancel all the claims to the species beyond five. He must subsequently file a divisional application, preferably prior to the issuance or abandonment of the parent application, in order to conserve the original filing date. The divisional application should disclose and claim the nonelected species of the parent case.

If the species claimed in the divisional application were patentable, the claims cannot be rejected as unpatentable over any of the species claimed in the parent application. On the other hand, if the species claimed in the divisional application are unpatentable, a requirement for divisional application cannot be effected.

9

Patent Interferences

WHAT IS AN INTERFERENCE?

An interference is a statutory proceeding instituted to determine priority of invention between two or more inventors. The main issue is the question as to who made the invention first. It has been reported that about 4 percent of all the patent applications are subject of interference proceedings, and as many as twenty inventions have been involved in one proceeding, although majority of the cases are instituted between two pending applications.

A proceeding for an interference may be declared: (1) between two or more pending applications, (2) between an issued patent and a pending application, and (3) between two issued patents. The first two cases are instituted in the Patent Office, whereas the third is instituted in federal courts, and the Patent Office has no jurisdiction over it.

INTERFERENCE PROCEEDINGS

In current practice, interferences between pending applications involve the allowable claims of rival contestants. For instance, when the examiner determines that the allowable claims of two different applicants read on the same invention and are interfering, he declares an interference. He sends the rival applicants copies of the allowable claims and requires that each applicant copy the interfering claim into his application if not already in it. Each rival applicant is also required to file a preliminary statement under oath by a certain date affixed by the Patent Office. This statement includes: (1) conception date, (2) dates of first drawing and written description, (3) dates of first disclosure to others, (4) date of reduction to practice, (5) date when the inventor began to perfect his invention, and (6) dates and serial numbers of foreign applications if filed prior to filing in the United States Patent Office. Thus it is readily apparent why it is very important to keep a complete record of inventions or developments in the daily log book furnished to each engineer and technician by the company where he is employed.

Interference practice is exceedingly complicated. Dates of conception and reduction to practice are generally relied upon in determining the outcome of an interference. It is not sufficient for an inventor to be the first to conceive an invention; he must also reduce it to practice within a reasonable time. Actual reduction to practice is not complete until the inventor builds a working model and tests it. Making sketches or blueprints is merely a step in the reduction to practice, which results only when a practical specimen is built and operated in the manner intended. Diligence in reduction to practice must be unabated and must start prior to the conception of the second party (who might have invented the same thing).

TYPES OF REDUCTION TO PRACTICE

There are two types of reduction to practice—actual and constructive. The first occurs when the inventor demonstrates the workability of a specimen of his invention. A constructive reduction to practice occurs when the inventor files an application in the Patent Office. However, it is not always the one who files his application first that is granted a patent. He must also show earlier dates of conception and reduction to practice, with continuous diligence (engineering improvements) thereafter until the invention was filed. As has been stated previously, it is a complicated procedure to prove priority.

When an interference is declared, the party filing his application first is the senior party, and the other is the junior party. The burden of proof is upon the junior party, whose chances of winning the case are estimated to be one out of eight. The following examples will illustrate a few of the many interference situations.

1. Smith and Jones conceived at the same time, but Smith reduced it to practice first. Smith wins.
2. Smith conceived in January and reduced to practice in November, using continuous diligence. Jones conceived in April and

reduced to practice in August. Smith wins. If no diligence is proved on the part of Smith, Jones wins the case.

3. Smith conceived after Jones, reduced to practice before Jones, and filed before Jones. Smith wins.
4. Smith conceived and reduced to practice first, but did nothing further. Then Jones conceived, filed an application, and received a patent. Thereafter, Smith filed an application. Smith loses because he concealed his invention until Jones received a patent.
5. Both Smith and Jones conceived and reduced to practice at the same time. Jones applied first. Jones wins, since no two patents can be issued on the same invention.

DETRIMENTS OF INTERFERENCE PRACTICE

There are two unfortunate features of interference practice: (1) dragnet application and (2) intimidating interference. The first involves an application filed by a party to keep a close watch on the developments in a certain art, for instance, in servo valves, for the sole purpose of forcing the Patent Office to declare an interference. As in all interference cases, the Patent Office has to disclose, to all parties concerned in the interference, other pending applications on servo valves. Such a party may abandon further action after receiving the information he sought. An intimidating interference involves the case in which an inventor, having a pending patent application, shows it to a second party for possible sale. If imprudent, the second party may file an application to cause an interference. If the inventor is frightened by the finances involved in such a proceeding, the second party reaps the fruit. Again, if the inventor has kept adequate records, such imprudence can be and has been defeated.

10

Infringement

WHAT IS INFRINGEMENT?

Practicing (making, using, and selling) a patented invention without the consent of the patent owner is infringement on the patent in the country where the patent is granted. The patent is ineffective in all other countries where the invention is not patented by the same patent owner. The manufacture of an article patented in this country for shipment and sale abroad is infringement. An article patented in the United States cannot be bought and manufactured abroad, by other than the patent owner, for sale in the United States. On the other hand, if the process of making an article, but not the article itself, is patented in the United States, it is not infringement to manufacture the article abroad and import it for sale in this country, nor is it infringement to make the same article in the United States by some other method.

WHEN IS INFRINGEMENT EFFECTIVE?

Infringement is effective only during the life (17 years) of a patent. No other time, before patent is issued or after it has expired, could any infringement be committed. To practice someone's invention while a patent on that invention is pending is not infringement. But, as soon as a patent is granted, the patentee can stop all unauthorized manufacturers, purchasers, and users from using or selling the article, even though the persons may have a number of produced articles on hand. They can neither use nor sell these articles without the consent of the patent owner. Hence, it is of utmost value to a manufacturer or purchaser to heed the markings "Patent Applied For" or "Patent Pending" on manufactured articles.

A peculiar infringement situation is buying the parts of a disassembled patented machine or item, assembling it without the patent owner's permission, and using or selling the machine or item after assembly. On the other hand, buying a patented machine already assembled, overhauling it, and then selling the assembled machine is not considered an infringement.

THE BREADTH AND SCOPE OF CLAIMS

A patent dominates an invention within the scope of its claims. A claim dominates an invention that contains every part, element, or step mentioned in that claim and nothing else. Each claim of an invention may be considered as a separate patent. A patent is infringed when any one of its claims is infringed. An invention dominated by a claim cannot be combined with other things, unless the claim is restricted in scope by containing limitations, which means that the claim describes a specific structure in rather narrow terms.

INFRINGEMENT REMEDY

When infringement occurs, the remedy is a suit brought in the federal court of the district wherein the rights of the patent owner have been violated. Since commercially successful articles are distributed nationwide, the infringement suit may be instituted in several districts at the same time. The infringers who are sued comprise the manufacturer, wholesalers, retailers, and innocent users. Any one or all of these may be enjoined from using the infringing article and can be sued for damages. When customers are thus sued, any reputable manufacturer must defend the suit to relieve them from liability. Consequently the dealer of a new machine or article might well investigate carefully the patent situation before entering into a transaction with the manufacturer of unknown repute.

LITIGATION

In a previous paragraph it was stated that infringement is the practice of a patented invention without the patent owner's consent in the country where the patent is granted. Litigation is a legal proceeding against the infringer and is an effective means of maintaining, for a period of years, a monopoly that was originally found on basic patents. Such a proceeding is determined by two principal factors: (1) the infringer must have money to pursue the case, and (2) enough infringement must have occurred to warrent a judgment which will pay more than the cost of the suit, unless the patentee's intention is only to stop the infringer from practicing his invention. The legal cost of an ordinary litigation is high and may amount to as much as $10,000.

If a number of companies have infringed a patent, the lawsuit should be brought against the weakest one first or to one to whom the invention is of little importance to defend the case. A court decision favorable to the patentee in such a case would establish a precedent that may influence decisions in succeeding proceedings.

An infringement suit is usually brought in equity. In such a suit, the court can award damages and an injunction against the infringer. Violation of this order will constitute contempt of court, and, if the infringer is found guilty, he may be penalized more stringently. Infringement suits are usually fought in the federal court, and jurors are rarely used, since their opinions on patent matters are considered less than nil.

LEGAL DEFENSES

An infringer who is a defendant in a litigation may advance three types of defense: (1) the defendant is not actually infringing the patent; (2) the person to whom the patent was granted is not the true and sole inventor (a fact seriously to be considered in applying for and being issued a patent); and (3) the infringement suit is barred by the Statutes of Limitations, meaning that the suit was not filed within the statutory six years after infringement was committed. The proof of infringement rests with the plaintiff (patentee or patent owner), whereas the proof of invalidity of the patent rests with the defendant (infringer). One or more patent claims will constitute infringement for damages, regardless of the remaining claims which may be declared invalid. When invalid claims are present in addition to valid claims in a patent, the patentee promptly must file a disclaimer or a reissue application to render his infringement suit effective. Otherwise, the whole patent may be held invalid.

AMOUNT OF DAMAGES

A definite formula for reckoning damages in infringement suits cannot be set up, since too many variables exist in the conditions surrounding infringement. However, an infringer's liability may amount to the definite loss suffered by the plaintiff due to competitive practice plus the profits made by the infringer from such an infringement. In cases where the infringement was deliberate, the judge may impose several times the normal damages. Triple damages are imposed when the plaintiff can prove fraud by the infringer.

PATENT INSURANCE

Insurance to guarantee innocent infringers against expense of infringement litigation may be had from Lloyds of London. Before entering into an insurance contract, a validity and infringement search is made on the patent so as to reduce the risks to a minimum. The patent buyer then is reasonably safe to practice the patent with an unknown background.

11

Patent, a Guide to Product Development

VALIDITY INVESTIGATION

All patents initially issued are presumably valid. That is, they are presumed to have been issued in accordance with the law defining the monopolies to which a patentee is entitled in the practice of his invention. Unfortunately, such a presumption is inconclusive, since the validity of a patent is not assured until after final adjudication. But, it would be impractical to have each granted patent adjudicated; such a practice would be expensive, time-consuming, and quite involved. An alternate method would be a thorough investigation of the state and history of the art by the applicant. This will give an idea as to the possible decision that may be expected in an infringement suit.

A validity investigation covers pertinent patents (domestic and foreign), printed publications (such as textbooks), trade magazines, catalogs, technical indexes, and newspapers. When the search results in a favorable finding, the patent or patents thus uncovered are valid, and it is considered a safe ground to avoid practicing the scope of such patents in an engineering development.

HOW PATENTS MAY BE DECLARED INVALID

Patents may be declared invalid for many reasons. Some of the grounds are: (1) patentee was not the first or actual inventor; (2) the subject of the patent was in public use for more than a year prior to the date of application; (3) the invention was described in a printed publication prior to the patentee's date of invention; and (4) the invention was set out in detail in a prior patent. It must be borne in mind that the use or sale of the invention by the inventor (or his assign) more than a year prior to the date of his application is sufficient ground to invalidate his patent when issued. In many countries, publication anywhere in the world immediately becomes a bar to the grant of a patent subsequently applied for.

Upon investigating the probable validity of a patent or patents, an inventor or his assigns will form an opinion as to the advisability of proceeding further with the development of the product under consideration.

VALIDITY RESULTS APPLIED TO PRODUCT DEVELOPMENT

An opinion on the subject of infringement based on the findings of patent and literature search is only as reliable as the person or persons making the search and signing the opinion. It is, therefore, a waste of time and money to obtain an opinion of a novice. By giving competent advice on possible infringement, a company patent department can do the most constructive work in helping to reduce creative ideas to practice. The company patent department seeks to keep the company out of trouble, to avoid patent litigations, and to settle incipient lawsuits. This department further attempts to build up the company's own patent posture by obtaining patents on improvements developed by its engineering and research personnel. In both of these endeavors, a close coordination between the flow of information to the engineer and proprietary protection on the product thus evolved must be maintained. The patent department of your company is in a position to give you such a service, your and your company's interests being its prime concern.

PATENT INDEX FILE

Some companies maintain a complete patent department in Washington, D.C. Patent work of the company is closely coordinated with the patent department, which executes all documentary and legal matters pertaining to patent practice. In addition to a Washington office, these companies may also maintain a patent file resulting from patent search on products pertinent to the company's operations and accumulated from additional surveys. The patents are retained and indexed, according to numerical order of patent numbers and alphabetical order of invention titles. This may be complemented by another cross-index file consisting of patent classification. The latter index is based on the patent office classification and contains all necessarily available patents, classified according to the arts in which they fall. Since it would be prohibitive to collect all classes of patents for any given art, the index may be classified under main classes, omitting subclasses. Many of the subclasses may contain matter which is an unimportant aggregation of the main class; in this manner, it may take thousands of volumes to complete all pertinent art on a given subject matter, and this procedure would be time-consuming and costly.

The classified index file will afford a ready reference to the available patents and will enable one to obtain information contained in these patents relating to any field of endeavor. Anyone working on a creative idea will want to obtain all available patents and study these carefully before finalizing the concept criteria of his new invention.

COMPLEMENTARY SEARCH

In some cases, a rough sketch together with a brief description may be necessary to make further patent search in the particular classification to which the idea pertains. A preferable procedure to follow would be the following:

1. Enter all records comprising sketches and description in the daily project log book as concepts become crystalized. Date this record.
2. Explain the workings of the idea to one or two witnesses and have them sign each page of this preliminary record in the project log book.
3. Prepare an Inventor's Record Form and have a patent search made by the patent department, with comments relating to the findings.
4. Study the patents uncovered in this search and, through assistance of your company patent service, determine a course of action. This action may involve proceeding with the original idea layout, improving on it by information gained from uncovered patents, or abandoning the idea because the invention has been fully anticipated.
5. In important cases where there is doubt of an anticipation, it is advisable to procure copies of the entire file wrapper of the

patent from the Patent Office. The actions, amendments, and arguments used will furnish valuable clues as to what to expect if an application is filed on the idea.

6. When an application for patent is warranted from these findings and is subsequently filed, all dates of conception, construction of the first working model, and similar dates should be kept secret until the patent is issued, and even thereafter.

12

Evaluating a Patent

FACTORS GOVERNING VALUE OF A PATENT

It is difficult to predict just how much a patent is worth and how much financial return it will bring to the patentee. A patent is worth as much as one can get for it. Its value, in general, will depend on the novelty, demand, and the ease of manufacturing. These requisites should be backed by a concrete sales organization. Likewise, the strength of a patent to exclude most of the competitors from the field will be another dominating factor for the successful commercialization of the patent. A patent with a claim or claims which can dominate all possible future improvements and modifications by a competitor would be obviously more valuable than one in which claims can be circumvented.

In spite of the fact that a patent may have been fortified by a number of broad, as well as narrow, claims does not warrant validity of such a patent. For instance, a patent may be declared invalid by a great number of reasons, some of which are:

1. The person in whose name the patent was granted was not the true or sole inventor.
2. The invention is not really an invention but an aggregation.
3. The invention is not novel; someone else has earlier invented and practiced it, but during the search and prosecution of the patent, this information was not brought to light.
4. The invention is public property because one of the bars to patentability, which states that the invention must not be described in a printed publication more than one year prior to the application, has been violated.

5. The patent does not describe the invention in full, clear, concise, and exact terms antecedent to the broad claims it contains.
6. New inventive matter has been introduced into the application during amendments.
7. The invention is not useful, or it does not belong to one of the classes of things that may be patented in the United States.

Thus, it is readily understandable why many prudent manufacturers are hesitant to purchase an outside invention which is presumably new and novel in its field. They deliberate, consult their attorneys, and even make additional market and patent searches before they are ready to "talk business." In line with this, a patent may have a much greater value in one person's hands than in another's. Generally, a patent is worth more in the hands of those who intend to exploit the invention directly than when it is for outright sale. A patent would be worthwhile for an individual who intends to manufacture and sell the article, rather than issue royalty licenses. The returns from royalty licenses may not pay the cost of patenting. Here, again, a man with a good invention can derive more from it when he can interest his employer in exploiting and commercializing the invention or patent. One will be reasonably safe in expecting that his employer will deal fairly with him and compensate him for his efforts in the creation of his invention. This statement should be strictly on the basis that the inventor or patentee has created the invention on his own time at his own expense, and that he is merely approaching his company or employer to "buy" the invention. In order to warrent a substantial outlay, the compensation will naturally depend on how "valuable" the invention is to the company and how much financial return the company will derive from it.

The value of a patent, then, is not a fixed, definite thing. No one knows how much a patent will earn before it has earned it. Sometimes this can be estimated with accuracy, but at other times the estimates are mere guesses. How, then, would one evaluate his patent?

METHODS OF EVALUATING A PATENT

The inventor may choose one of several different methods of profiting from his invention. The evaluation of the worth of his patent will depend on which method he chooses to commercialize his invention. Some of the common methods of practicing an invention follow.

Practicing of Invention by the Inventor

An inventor may practice his invention himself. Although this involves the greatest amount of work with the greatest risk, the possibility of the greatest profit to the inventor may warrant this method, if the inventor possesses a moderate amount of capital and management ability. This will give the inventor not only what a possible buyer or licensee would pay him, but also what the latter would make. A moderate success in a self-owned business will greatly improve the chances of later disposing of the patent to the prospective buyer on terms favorable to the inventor.

In majority of the cases, the inventor does not have the facility and the capital to manufacture and sell his invention. In such a case, he may try to interest investors, who may organize the business and commercialize the invention under the inventor's technical direction. Or, the inventor may contract with an established organization to distribute and sell his invention on a contract basis, in which case the inventor is the manufacturer of the product.

In either case, the inventor may derive the highest profit, if any profit is to be made on the invention, by remembering that the product must have demand and thus be salable. Again, until the patent is exploited, no monetary value can be attached to the invention.

Selling the Patent Outright

The inventor may choose to sell his invention for a cash consideration. Although the sale of the invention may bring immediate cash, the inventor seldom profits very much from such a transaction. Since the buyer is taking the risk of a cash royalty, his offer will be limited to an amount which he might lose without financial inconvenience in case the invention does not prove to be a commercial success. To alleviate this situation for both the inventor and the buyer, the inventor may take a small down payment, followed by royalties when the invention is placed on the market. Such an arrangement will be easier on the buyer and more profitable to the inventor if the product proves successful.

Selling the Patent on a Royalty Basis

Most buyers will try to buy an invention on royalty. This has the advantage that the buyer does not have to put up an outlay of cash in one lump sum, since he cannot deduct this outlay of money from his tax returns until the invention has been practiced and profit has been made. The inventor may find a buyer much easier on such a term than on an outright sale basis. Thus, if the inventor can wait until royalties are derived, which may take from several months to several years, his ultimate income or profit is likely to be more.

Licensing the Patent

A patent can be licensed to more than one party, unless an arrangement is made to license exclusively. When a patent is licensed to more than one party, the chances of deriving an equitable amount of income from the invention increase. If one party fails to make a success in his exploitation of the invention, and the next one is successful, the inventor will be deriving uninterrupted income from his invention.

Doing Nothing With the Patent

After issuance of a patent, some inventors lay the patent aside and do nothing about it. Of course, unless diligence is used to exploit a patent, no profit can be expected from it. One inherent danger of such a procrastinated action is that as the years go by without commercializing the patent, either another person may come up with something similar or the patent will become older and its sale value will suffer.

Depreciation of a Patent

An uncommercialized patent can be depreciated and tax deductible after 17 years, the useful life of the patent. Depreciation is allowed from the date of the granting of the patent and is computed on its entire cost. The cost will include the government fees, attorney's fees, drawings, experimental models, and other developmental and experimental or testing costs paid by the patentee. All such costs must be bona fide, and evidence must be available, as for any other tax deductible item.

13

Copyrights

TWO COPYRIGHT RIGHTS

A copyright is the right granted to a person to prevent others from copying an intellectual creation. The right includes preventing others from copying one's unpublished work; this is common-law copyright. The second, called the statutory copyright, is the right to prevent others from copying the published work. A copyright issued by the Library of Congress includes both, although it is commonly known as a statutory copyright.

Common-Law Copyright

The right of the creator of an unpublished work arises from the common-law right of a man to privacy and protection against breach of confidence. For instance, a writer who sends an unpublished manuscript to a publisher has the right to sue the publisher for damages if the latter publishes the manuscript without the consent of the writer, even though the writer has not been granted a copyright. However, the publisher may in turn rewrite the manuscript and publish what he has written, unless he has specifically agreed not to publish anything similar to the contents of the manuscript.

Statutory Copyright

The common-law copyright protects an intellectual creation from the time it is created until it is published, when the common-law protection ceases. But, if the work is registered in the Library of Congress, a statutory copyright is effected for the useful life of 28 years. The copyright may be renewed for an additional term of 28 years, making a total of 56 years for the work. Thereafter the work becomes public property, so that any one can use it without due concern to the originator.

The statutory copyright may be obtained by sending two copies of the published work, containing a copyright notice and the copyright fee, to the Register of Copyrights, Library of Congress, Washington, DC 20540. The fee for registering a simple case is $6.

WHAT MAY BE COPYRIGHTED?

The Copyright Office does not grant copyrights but merely registers claims to copyrights when statutory requirements for such registering are complied with. In applying for the registration of claims to copyrights, any one of the following forms furnished by the Office may be used:

Class A: Form A–Book published in the United States.
Class B: Form B–Periodical.
Form B5–Contribution to a periodical.
Class C: Form C–Lecture, sermon, address, prepared for oral delivery.
Class D: Form D–Dramatic or dramatic-musical composition.
Class E: Form E–Musical composition.
Class F: Form F–Map.
Class G: Form G–Work of art; model or design for work of art.
Class H: Form H–Reproduction of work of art.
Class I: Form I–Drawing or plastic work of a scientific or technical character.
Class J: Form J–Photograph.
Class K: Form K–Print and pictorial illustration.
Form KK–Print or label for article of merchandise.
Class L: Form L–Motion-picture photoplay.
Class M: Form M–Motion picture other than photoplay.

WHO MAY ACQUIRE COPYRIGHT?

A copyright may be issued to the originator of a work, his assigns, or administrators. Two or more persons as joint authors may receive a copyright if they have participated in creating the work. The originator must be a United States citizen, or he must have resided in the United States at the time the work was first published.

COPYRIGHT NOTICE

This notice consists of three elements: the word "Copyright," the year date, and the name of the copyright owner. For example, John Smith has copyrighted a book in May of 1972. The notice in the book will read, "Copyright 1972, by John Smith." Another way of expressing this notice will be to use the capital letter "C" of Copyright followed by the year date and the author's name or the copyright owner's name (such as a publisher). It will be written "© 1972, by John Smith." Most U.S. publishers also add the words, "All rights reserved." This is a requirement of countries belonging to the Buenos Aires Convention (most Latin American countries and the U.S.).

In a book, the notice usually appears on the reverse side of the title page; for a periodical, the notice appears on the title page under the title heading. For a musical work, the notice appears on the title page or on the first page of the music. For the copyrighted instructions of a game board entitled, for instance, 3-Ina-Ro Educational Game, the notice will take the form:

3-INA-RO EDUCATIONAL GAME
Copyright 1973
By
Scientific Enterprises, Inc.
Los Angeles, California 90019

If the copyright notice is placed anywhere other than on the title page, reverse side of the title page, or on the first page of a piece of sheet music, it will have no effect. Also, if the copyright notice is attached to the work after it is published, the notice will have no effect, and the owner will have forfeited his right to a copyright. Any unpublished work should bear the copyright notice if the work is to leave the owner's control; this avoids the danger of inadvertent publication by someone else.

14

Registration of Trademarks and Trade Names

Strictly speaking, a trademark is a mark or name used on a vendible commodity, whereas the term "trade name" refers to an entire business and its good will. For instance, Robertshaw is the trade name denoting the goodwill of a company, as well as its entire business, including all lines of products it manufactures. The word CRYST-OVEN may be the trademark used only for all crystal ovens built by the Robertshaw-Fulton Controls Company.

The primary function of a trademark is to indicate origin. A trademark also serves to guarantee the quality of the goods through advertising and to maintain a demand for the product. Trademarks are applied to goods as well as to services and associations. Rights in a trademark are acquired only by use, which must continue to preserve the rights. A trademark right is perpetual as long as it maintains its function of distinguishing one manufacturer's goods from that of the owner of the trademark. Trademarks are registered in the patent office for an indefinite period.

Trademarks are not protected by the patent or copyright laws. The right to a trademark must be created by the owner through commercial use of the mark on his goods. The mere fact that a person designates or coins a clever trademark does not give him trademark rights to that mark. The mark must be original, the first one of its kind, and must be in use in interstate commerce or abroad.

The Trademark Act of 1946 provides two registers, designated as the Principal Register and the Supplemental Register. Coined, arbitrary, fanciful, or suggestive marks, are generally referred to as "Technical Marks," and may be registered on the Principal Register.

Marks not qualified for registration on the Principal Register but which, nevertheless, are capable of distinguishing applicant's goods and have been in lawful use in commerce for at least one year may be registered on a Supplemental Register. A technical mark must be in commercial use for the five years preceding the date of filing of application for registration.

A trademark registered under Principal Register is contestable within 30 days of its publication in the Official Gazette, whereas a mark registered under Supplemental Register is not contestable.

When a mark meets (after a careful preliminary search) the requirements of a good technical trademark and has been used in interstate or foreign commerce, its owner can register it in the United States Patent Office, so that no one else can use the same mark on his goods. Registration is prima facie proof of ownership of the mark. The owner of the mark can sue infringers in the federal courts. Registration entitles the owner to mark his goods "Reg. U.S. Pat. Off.," which tends to discourage infringement.

The owner may file and prosecute his own application for registration, or he may be represented by an attorney. The application consists of a short statement indicating the name and the first date of use of the mark, a drawing of the mark, five specimens or facsimiles, and the required fee of $35.

The owner of the registered mark must file an affidavit within 6 years after the mark is registered, showing that the mark is still in use. Otherwise, the registration will be cancelled. The registration is good for 20 years, and it may be renewed within 6 months before its expiration, for a fee of $25 paid to the Patent Office.

Appendix

This appendix contains a sample of a utility patent (pages 82 to 86), a design patent (page 87), a trade mark (page 88) and drawings and symbols for draftsmen (pages 89 to 92). The facsimiles are only for illustration and should not be construed as the best example of the class described. The utility patent, because of the many prior disclosures, contains rather specific claims; no generic claim is shown nor could be drafted. However, it offers sufficient protection against imprudent use of the construction as shown and claimed. The design patent is a typical example and the form can be used for any other design patent application. When color is a distinguishing feature of the invention, the pertinent parts may be shaded in accordance with the hatched lines shown at the bottom of the page on Symbols for Draftsmen. Additional instructions for other symbols are self-explanatory.

UTILITY PATENT

May 19, 1970 W. L. BURLEY ET AL 3,513,290

AUTOMATICALLY HEATED ICE-CREAM SCOOP WITH STAND

Filed April 20, 1967 2 Sheets-Sheet 1

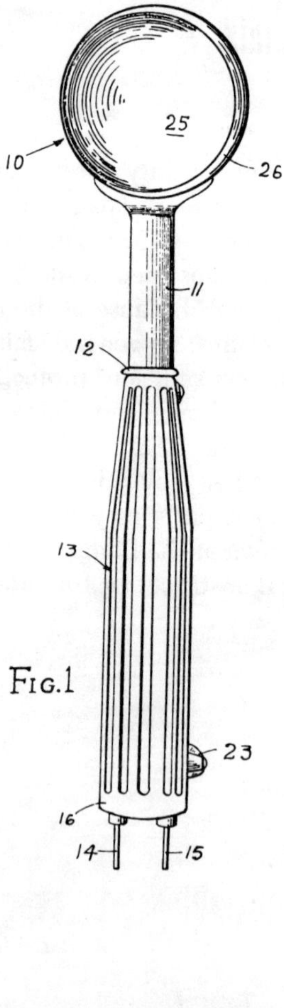

FIG. 1

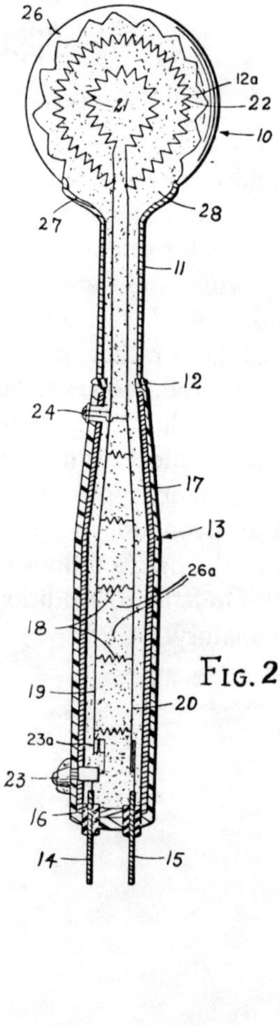

FIG. 2

May 19, 1970 W. L. BURLEY ET AL **3,513,290**

AUTOMATICALLY HEATED ICE-CREAM SCOOP WITH STAND

Filed April 20, 1967 2 Sheets-Sheet 2

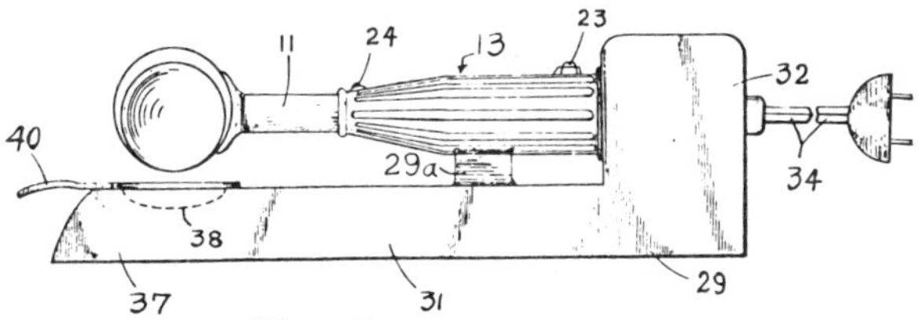

FIG. 3

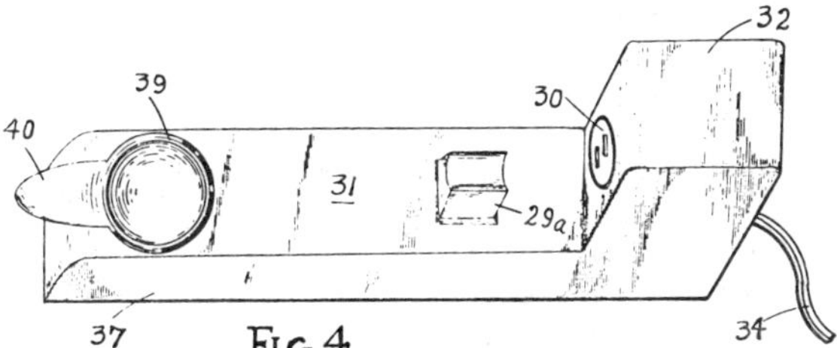

FIG. 4

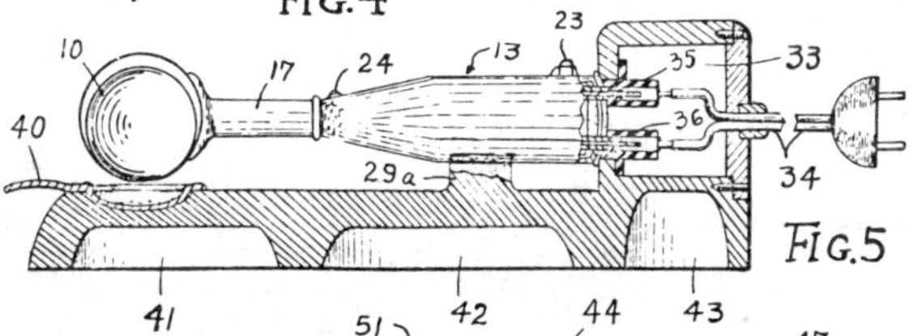

FIG. 5

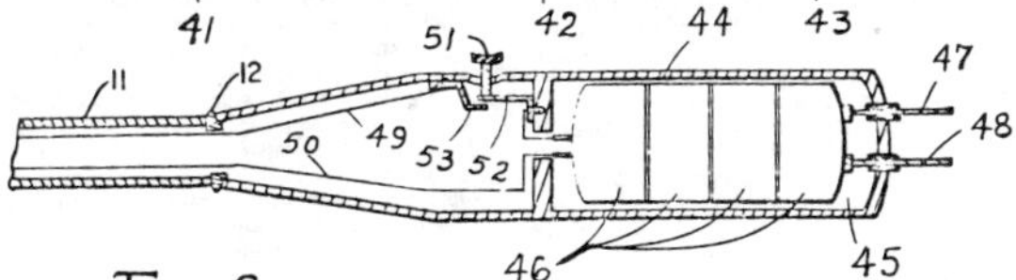

FIG. 6

United States Patent Office 3,513,290
Patented May 19, 1970

1

3,513,290
AUTOMATICALLY HEATED ICE-CREAM SCOOP WITH STAND

Wallace L. Burley, Santa Ana, Calif. (Rte. 3, Box 141, Troutdale, Oreg. 97060), and Hrand M. Muncheryan, 1135 N. Morningside, Orange, Calif. 92667

Filed Apr. 20, 1967, Ser. No. 637,862
Int. Cl. A47j *43/28;* H05b *1/02*
U.S. Cl. 219—242 8 Claims

ABSTRACT OF THE DISCLOSURE

A cordless electrically heated ice-cream scoop is provided with a stand to horizontally support the scoop when not in use. The stand includes a receptacle for catching ice-cream drippings from the scoop. In one embodiment, the handle and the ladle of the scoop are provided with electric heating elements surrounded by a filling of granular heat-retaining electrical insulation material such as diatamaceous earth or mica flakes. The stand incorporates an electric outlet cooperating with terminals in the handle to transmit current to the heating element. A thermostat is provided in the handle to maintain the scoop at a desired temperature. In a second embodiment, only the ladle is electrically heated and the handle contains rechargeable batteries for supplying current to the heating means. The stand includes electric outlet means for recharging the batteries.

The present invention relates to ice-cream scoops and more particularly to an automatically heated ice-cream scoop provided with a detachable stand having a power outlet and means for catching the drippings of ice cream from said scoop when not in use and placed on said stand.

Conventional ice-cream scoops presently used in soda fountains and homes usually offer difficulty in discharging the ice cream after the scoop has been dipped several times into the bulk ice cream. The temperature of the scoop is so lowered that the ice cream freezes therein and sticks to the walls of the scoop. In some scoops, mechanical scraping blades are provided to discharge the ice cream from the scoop, but they also freeze and become difficult to operate and, at times, they break, defeating the purpose of such attachments. Other scoops are also available with heating elements or heated fluids such as oil, liquid wax, mercury and the like, but these also are unsatisfactory in operation because of leakage of the fluids from wear of the device at the seals, contaminating the ice cream. As for the use of mercury as the heat sustaining element, which is very fluid and has low vapor pressure, when it leaks out and mixes with ice cream, it will cause dangerous consequences because of its poisonous character.

Overcoming these difficulties and deleterious effects of liquid-filled scoops, the present invention is designed to eliminate these troublesome problems by offering a simple-construction, efficient, durable, and practicable device for scooping and dispensing ice cream. The invention is provided with an adjustable temperature control to sustain the ladle of the scoop at a relatively uniform temperature. This function is further aided by filling the handle of the scoop and the space intermediate to the heating elements with a thermally conductive and electrically insulating inert substance such as flakes of mica or granules of diatomaceous earth, which cannot be spilled or leak out of the scoop and further is not harmful as mercury or mineral oil stated in earlier patents. Furthermore, the invention is provided with a stand which serves to support the scoop in a horizontal position to prevent the ice-cream drippings from the ladle from reaching the electric socket, causing a dangerous short circuit, as would be prevalent

2

in any stand which supports the ice-cream scoop in a vertical position. In addition, the stand receives and transmits heating current to the scoop from an ordinary 115-volt household power source.

The principal object of the invention is to provide an ice-cream scoop which is automatically heated to a predetermined temperature and is adapted to retain that temperature during use, thus facilitating the discharge of the ice-cream from the scoop.

A further object of the invention is to provide an automatic control means for adjusting the current which heats the scoop and thereby sustaining the temperature constant within reasonable tolerances.

Another object of the invention is to provide a thermoconductive current-insulating solid matter within the scoop capable of heating and retaining heat for long periods of time.

A further object of the invention is to provide a stand having incorporated therein an electrical outlet means which serves to furnish current to the heating elements of the scoop as well as to support the scoop in a horizontal position when not in use, whereby the ice-cream drippings therefrom can be collected in a removable receptacle provided in the stand.

A still further object in an alternate embodiment of the invention is to provide a scoop of the character having a self-contained electrical heating source which can be recharged from an ordinary 115-volt household current after its power is consumed in the heating elements.

Another object in the alternate embodiment of the invention is to provide a pushbutton switch on the scoop having self-contained current source in which the current can be turned on and off as desired for heating the ice-cream scoop.

Other objects and advantages of the invention will become better understood hereinafter from a consideration of the specification with reference to the accompanying drawings forming part thereof, and in which like numerals correspond to like parts throughout the several views of the invention, and wherein:

FIG. 1 is a front view of the invention in elevation,

FIG. 2 is a vertical sectional view of the ice-cream scoop, with a cutaway section showing the back side of the scoop, the heating elements, and the related parts.

FIG. 3 is a side perspective view of the ice-cream scoop mounted in normal resting position in the stand provided with a current outlet box,

FIG. 4 is a top view of the stand, with the scoop removed to show the electrical outlet plug and the receptacle for ice-cream drippings,

FIG. 5 shows the ice-cream scoop with the stand in the same position as that in FIG. 3, but with a sectional portion to show details of the power outlet box, base construction of the stand, and the removable receptacle, and

FIG. 6 is a modified embodiment of the invention in longitudinal sectional view, showing the arrangement of the self-contained power supply and the current control switch, the ice-cream scooping portion being the same as that shown in FIGS. 1 and 2 and left out for space restriction on the drawing.

Referring to the drawing, the improved ice-cream scoop characterizing the present invention comprises a concave double-walled hemispherical member or scoop section **10** for scooping and dispensing ice cream and having a tubular extension **11** terminating in an annular flange **12** which is secured to a supporting grip section or handle **13**. Electrical prongs **14** and **15** are attached to the terminal end **16** of said handle **13** to receive and supply current thereto. A chamber **12*a*** is formed in said double-walled scoop section **10** and is continuous with chamber **17** formed within the tubular section **11** and the handle **13**, wherein is disposed a plurality of heating elements **18** connected in parallel in the circuit formed by conductors **19** and **20** supplied with electric current through prongs **14** and **15**. The conductors **19** and **20** are continuous with additional heating elements **21** and **22** located in the hemispherical member or scoop **10**. A variable thermostatic control **23** is connected in the circuit of the conductor **19** (or **20**) to adjust the current therein and thereby the temperature of chambers **17** and **12a**. The numeral **23a** is the thermosensitive element of the thermostatic control **23**. When the current is on, a red light from lamp **24** is on, indicating the scoop is heating; when the scoop **10** is heated to a temperature set by the variable-temperature control thermostat **23**, the light of lamp **24** is off. Thus the scoop is continuously energized as long as it is in the stand except when the thermostatic element **23a** opens the circuit therein.

The conductors **19** and **20** and the heating elements **18**, **21**, and **22** are covered with an electric insulator, such as asbestos. The heating elements **21** and **22** are sandwiched between a front cup **25** and an outer or rear cup **26** of the scoop **10** and are electrically insulated therefrom by their insulating sheaths and by means of the thermoconductive insulation material **26a** surrounding the heating elements. The peripheries of the cups **25** and **26** are sealed as by welding or by other suitable means; the cups **25** and **26** are further welded to a supporting conical flange **27** of tubular extension **11**. The space between the electric heating elements and the surrounding walls is filled with a thermoconductive and insulating material **26a**, such as granules of diatomaceous earth and flakes of mica or similar material, to retain the heat evolved within the chambers **17** and **28**. The electrically insulating thermoconductive material **26a** is represented by dots in FIG. 2.

When the ice-cream scoop is not in use, it is normally placed on its rack **29**, by inserting the electrical outlet prongs **14** and **15** into the outlet socket **30** (FIG. 4) provided in the rack **29** and supporting on a saddlelike projection **29a** formed thereon. The rack **29** comprises a flat extended section **31**, adapted to be supported on a horizontal surface, and a boxlike electric outlet section **32** containing a chamber **33** (FIG. 5) wherein an electric outlet cord **34** makes connections to split bus bars **35** and **36**, which form the contacts of socket **30** and provide mechanical support to prongs **14** and **15**. The flat extended section **31** contains at its terminal portion **37** a circular depression **38** wherein is disposed a removable pan or receptacle **39** for catching ice-cream drippings from the hemispherical scoop member **10** of the ice-cream scoop.

The pan **39** is provided with an elongated flange **40** used for holding the pan during disposing of the ice-cream drippings, or during washing the pan. By supporting the ice-cream scoop in a horizontal position as shown in FIG. 3 or FIG. 5, the handle **13** of the scoop and its associated parts do not become contaminated with ice-cream drippings, which also may short-circuit the current in the event the drippings reach the outlet contacts or bus bars **35** or **36** of the current source were the ice-cream scoop held in a vertical position in an electric socket. The openings **41**, **42**, and **43** in the undersurface of rack **29** are for reducing the weight thereof and saving material of construction.

In operation, the scoop is placed on the stand or rack **29**, with the prongs **14** and **15** inserted into the outlet socket **30**. The alternating current from the external source enters the scoop through the prongs **14** and **15**, and begins to heat the elements **18**, **20**, and **21**. The indicating light **24** comes on and remains "on" until the ladle part **26** of the scoop heats, whereupon the light of indicator **24** goes off. The ice-cream scoop is now ready for scooping ice cream.

In the modified embodiment shown in FIG. 6, the invention comprises a handle **44** having therein a chamber **45** containing rechargeable batteries **46** provided with contact prongs **47** and **48** to receive current from an outside source, such as from a 115-volt household circuit; the household current usually being alternating, it is rectified before it is applied to recharge the batteries. The batteries **46** are preferably connected in parallel and furnish heating current through conductors **49** and **50** to the heating elements (not shown). The conductors **49** and **50** are similar to elements **21** and **22**. A cutoff switch **51** is connected in series with the conductors **49** and **50** and is used to turn the current on as needed, by pressing thereon. The switch **51** is normally at open position and when depressed the resilient contact **52** makes an electrical contact with its mating contact stub **53**, which action turns on the current. When not in use, the scoop is supported on the rack **29** in the same manner as the scoop illustrated in FIG. 3. This action causes the batteries **46** to recharge if their charge has been consumed to a level below their normal charge. When the batteries **46** are fully charged, they are automatically cut off from the external current source; the mechanism of the automatic current cutoff is not shown because the commercially sold rechargeable batteries are provided with such a mechanism.

Having thus described the invention, it is to be understood that certain modifications in the construction and arrangement of the parts thereof will be made, as deemed necessary, without departing from the scope of the appended claims.

We claim:

1. An automatically heated ice-cream scoop with stand, comprising the combination of an elongated hollow handle having at one end thereof a hollow concave double-walled hemispherical means adapted to receive and dispense ice cream and at the other end electric conductor means for detachable connection to a source of electric current, electric heating elements positioned axially in said elongated handle and extending into the space within said hollow concave double-walled hemispherical means, means disposed in said handle for receiving and supplying electric current to said heating elements to heat said double-walled hemispherical means, a series-connected means in the circuit of said electric heating elements to control said electric current to said elements, and a stand means comprising a horizontal, flat section with an electric current outlet box disposed at one end thereof having electric inlet and outlet means, said inlet means being connected to an external source of current and said outlet means being detachably connected to said electric conductor means for transmitting current thereto; a continuous chamber formed within said handle and extending into said hollow concave double-walled hemispherical means and an electrically insulating thermoconductive material being disposed therein to fill the spaces surrounding said electric heater elements; said flat section of said stand means having thereon first means adapted to support said ice-cream scoop in a horizontal position and in parallel relation to said flat section of said stand means, and a second means disposed in said flat section for collecting ice-cream drippings from said double-walled hemispherical means when said scoop is positioned horizontally on said stand means with said electric conductor means on said handle of said scoop inserted into said outlet means in said electric current outlet box.

2. An automatically heated ice-cream scoop with stand as defined in claim 1, wherein said electric conductor means comprises a pair of metal prongs secured to the end of said handle, said means in said handle for receiving and supplying electric current to said heating elements comprising insulated electric wires extending from said prongs into said continuous chamber in said handle and connected to said electric heating elements to supply heating current thereto.

3. An automatically heated ice-cream scoop with stand as defined in claim 1, wherein said means disposed in

3,513,290

5

said handle for receiving and supplying current to said electric heating elements is a pair of electric conductors provided with said electric heating elements therebetween and connected in parallel thereto.

4. An automatically heated ice-cream scoop with stand as defined in claim **1**, wherein said series-connected means is a variable temperature-control thermostat for maintaining the temperature of said ice-cream scoop at a predetermined level.

5. An automatically heated ice-cream scoop as defined in claim **1**, wherein said series connected means is a thermostat disposed in the circuit of said heating elements and is further connected in series with an indicator lamp disposed in the wall of said handle of said scoop, said thermostat opens the circuit of said heating elements and turns off the light of the indicator lamp when the scoop is heated to the temperature set by said thermostat, thus indicating that the ice-cream scoop is ready for the next operation.

6. An automatically heated ice-cream scoop with stand as defined in claim **1**, wherein said first means on the flat section having the electric current outlet box is a saddle-like projection formed thereon in remote relation to said electric current outlet box, said saddle-like projection being provided with means to accommodate the handle of said scoop and to support said double-walled hemispherical means of said scoop in spaced and superior relation to said second means for collecting ice-cream drippings from said double-walled hemispherical means.

7. An automatically heated ice-cream scoop with stand as defined in claim **1**, wherein said flat section of said stand means is provided adjacent the terminal portion of said flat section with a circular depression formed in the upper surface of said flat section, and wherein said second means for collecting ice-cream drippings is a receptacle adapted to be accommodated in said circular depression in removable relation thereto, said receptacle having a handle for removal of said receptacle from said circular depression.

8. An article of manufacture, comprising in combination an ice-cream scoop, a stand for holding said scoop in a horizontal position, and an electric outlet box having an electric cord to transmit current thereto from an external source being formed at one end of said stand; said ice-cream scoop having a double-walled ladle and an elongated handle secured to said ladle for manipula-

6

tion thereof; said handle having a chamber axially formed therein; electric heating elements disposed within said chamber and extending into the space within said double-walled ladle and to the terminal end of said elongated handle, which latter end being closed and a pair of electric inlet prongs being secured therein and connected to said heating elements for transmitting heating current thereto; thermosensitive means connected in the circuit of said heating elements to control the current therein and thereby the temperature of said double-walled ladle; and, an electrically insulating thermoconductive heat storage material being disposed around the heating elements within said scoop and in said handle; said stand having a flat base section adapted to be placed on a horizontal surface and said electric outlet box formed at one end of said stand having an electric outlet socket means positioned at right angles to the top surface of said base section and electrically connected to the circuit of said electric cord; said socket means being adapted to accommodate said inlet prongs at the end of said elongated handle to transmit to said prongs electric current from said electric cord; said flat base section having a support means formed thereon for holding said scoop in a horizontal position when not in use, and a concave area formed adjacent the terminal end of said base section with a removable pan disposed therein to catch the drippings of ice cream from said scoop when mounted on said support means with said inlet prongs inserted into said electric outlet socket.

References Cited

UNITED STATES PATENTS

1,323,523	12/1919	Cox	107—48
1,971,577	8/1934	Parker	107—48
2,166,810	7/1939	Gammeter	219—242
2,256,770	9/1941	Armstrong	107—48
2,323,296	7/1943	Brown	248—37.3
3,293,528	12/1966	Rosen et al.	320—2
3,419,704	12/1968	Hunt	219—240 X

ANTHONY BARTIS, Primary Examiner

U.S. Cl. X.R.

107—48; 219—227, 240, 241; 248—37.3; 320—2

DESIGN PATENT

United States Patent Office

Des. 174,052
Patented Feb. 15, 1955

174,052

COMBINED PENHOLDER AND THERMOMETER

Hrand M. Muncheryan, Duarte, Calif.

Application May 17, 1954, Serial No. 30,538

Term of patent 14 years

(Cl. D52—7)

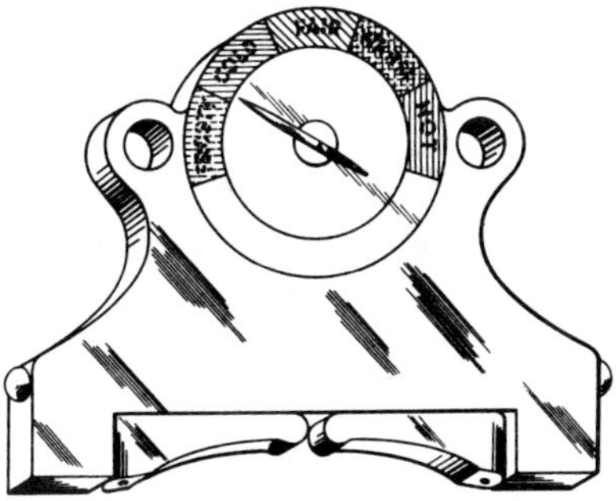

Fig. 1

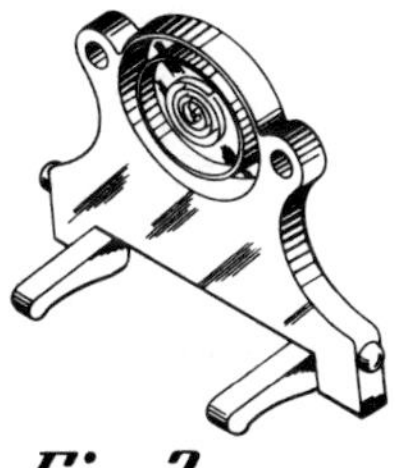

Fig. 2

To all whom it may concern:

Be it known that I, Hrand M. Muncheryan, a citizen of the United States and a resident of Duarte, State of California, have invented a new, original, and ornamental Design for a Combined Penholder and Thermometer, of which the following is a specification, reference being had to the accompanying drawing, forming a part hereof.

Figure 1 is a front perspective view of a combined penholder and thermometer showing my new design, and

Figure 2 is a rear perspective view thereof on reduced scale as it appears with the legs extended rearwardly.

The drawing has been lined for color.

I claim:

The ornamental design for a combined penholder and thermometer, as shown and described.

References Cited in the file of this patent

UNITED STATES PATENTS

D. 86,838	Hoffman	Apr. 26, 1932
D. 136,108	Casey	Aug. 3, 1943
D. 149,335	Beatty	Apr. 20, 1948
D. 160,077	Eaves	Sept. 12, 1950
D. 168,105	Mitchell et al.	Nov. 4, 1952
D. 171,653	Kuhlman	Mar. 9, 1954

TRADE MARK

Registered June 18, 1946 | **Trade-Mark 421,908**

UNITED STATES PATENT OFFICE

Scientific Enterprises, Inc., Los Angeles, Calif.

Act of March 19, 1920

Application October 15, 1945, Serial No. 489,989

ATOMIK

STATEMENT

To the Commissioner of Patents:

Scientific Enterprises, Inc., a corporation duly organized under the laws of the State of California and located at Los Angeles, California, and doing business at 5067 West Washington Boulevard, Los Angeles, California, has adopted and used the trade-mark shown in the accompanying drawing, for a GAME EMBODYING A FOLDABLE GAME-BOARD MADE OF CARDBOARD AND HAVING ON IT DIAGRAMS OF VARIOUS ATOMIC STRUCTURES SHOWING THE ELECTRON CONFIGURATION IN ATOMS, AND IS PLAYED WITH MARBLES, PLASTIC DISCS, OR PAWNS, in Class 22, Games, toys, and sporting goods, and presents herewith five specimens showing the trade-mark as actually used by applicant upon the goods, and requests that the same be registered in the United States Patent Office in accordance with the act of March 19, 1920. The trade-mark has been continuously used and applied to said goods in applicant's business and in the business of applicant's predecessors since March, 1938. The mark has been in bona fide use for not less than one year in interstate commerce by the applicant (or applicant's predecessors in business). The trade-mark is applied or affixed to the goods, or the packages containing the same, by printing said trade-mark thereon, or by placing thereon a printed label on which the trade-mark is shown.

SCIENTIFIC ENTERPRISES, INC.,
By HRAND M. MUNCHERYAN,
President.

DRAWINGS AND SYMBOLS FOR DRAFTSMEN

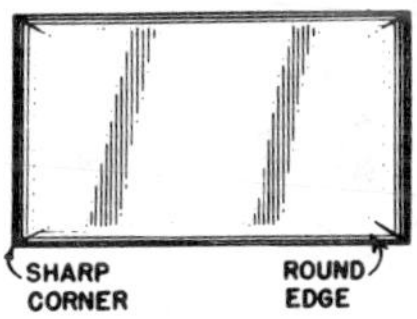

Irregular surface.

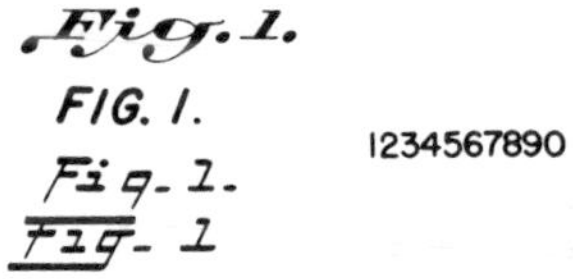

Some lettering styles used on drawing.

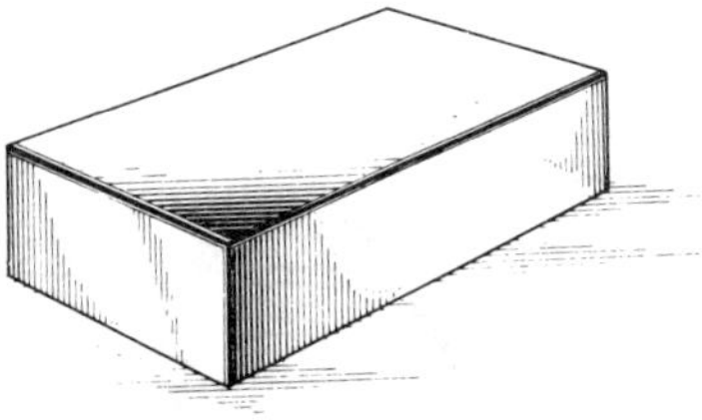

Perspective view of a rectangular object.

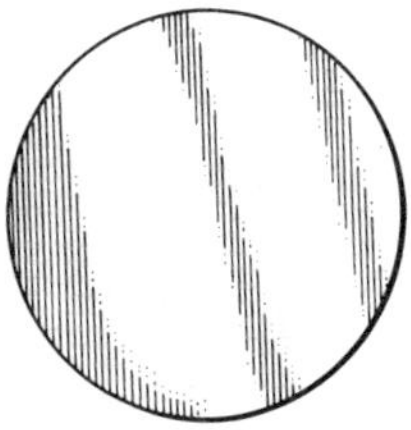

Surface shading for top view of a disk.

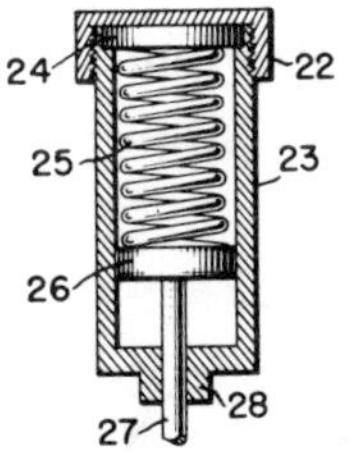

View showing the manner of representing a spring member.

Shadings for round handles and screw threads.

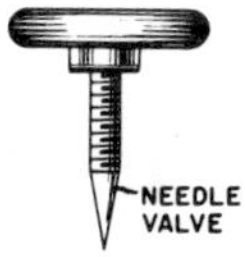

Note screw-thread representations.

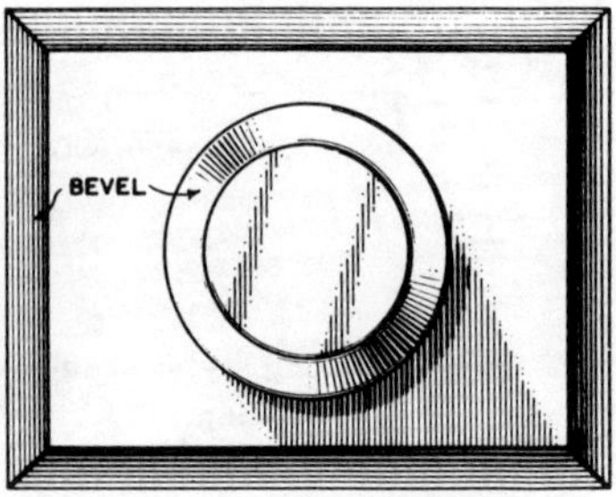

View illustrating surface shading.

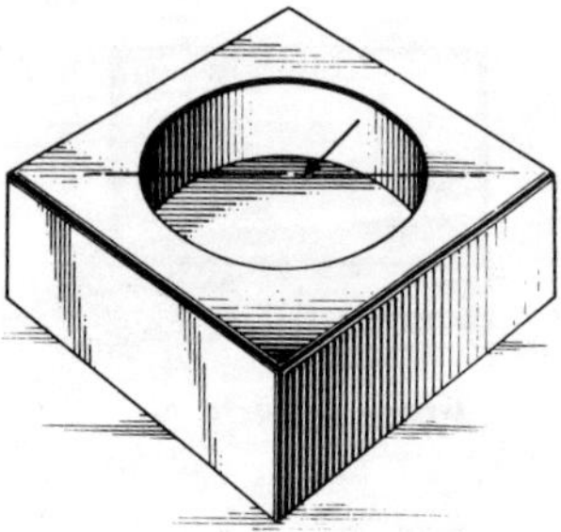

View illustrating the shading of a long axis of an ellipse.

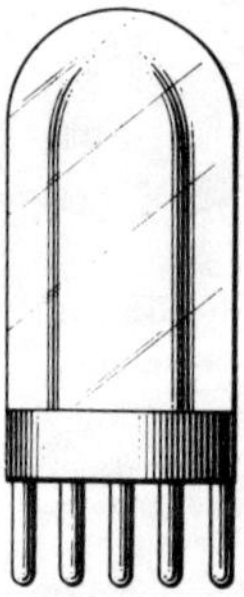

View illustrating the shading of the glass envelope of a radio tube.

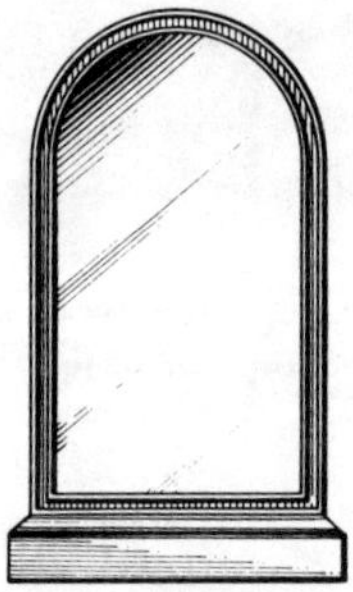

View illustrating the surface shading of a mirror.

Conventional shading for cylindrical parts.

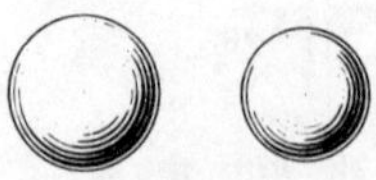

View illustrating surface shading for spherical parts.

Two spur gears in mesh.

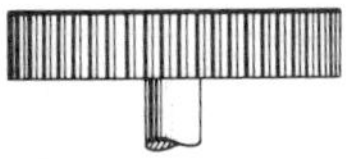

Spur gear.

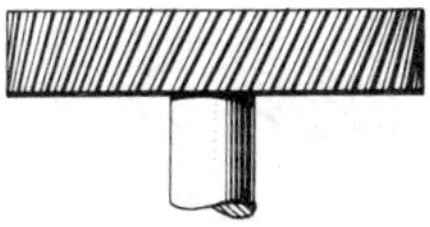

Helical gear.

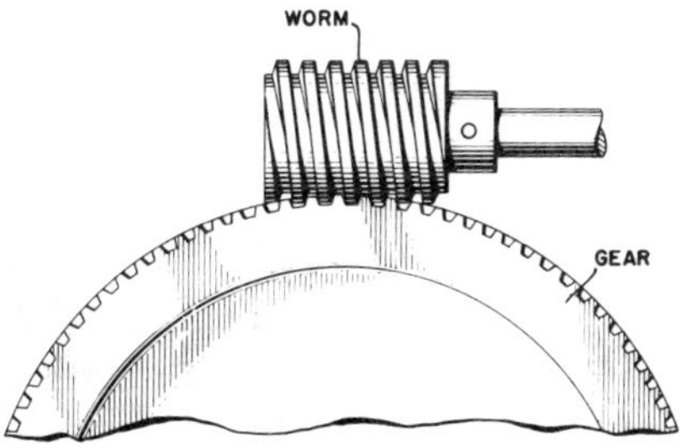

A worm gear in mesh with a worm. (No legend should be added, such as "GEAR" in the drawing.)

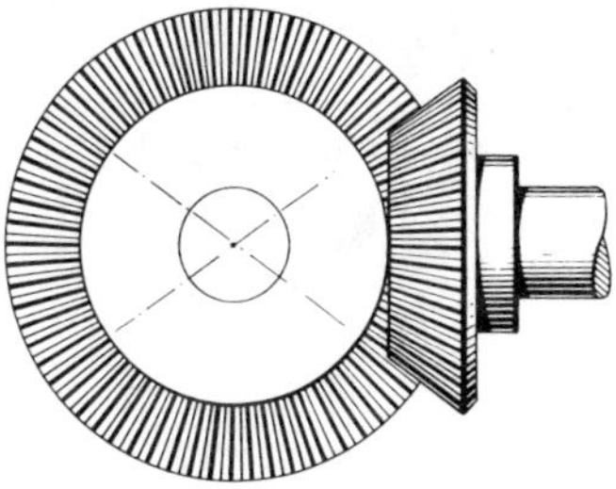

Top plan view of bevel gears. (The broken lines are only for instruction.)

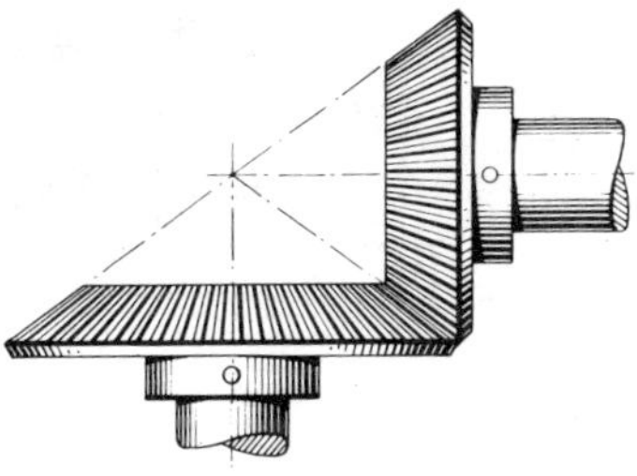

Bevel gears—teeth of each gear have the same slant.

SYMBOLS FOR DRAFTSMEN

Rule 84 (g) states that graphical symbols for conventional elements may be used on the drawing when appropriate, subject to approval by the Office. The symbols and other conventional devices which follow have been and are approved for such use. This collection does not purport to be exhaustive, other standard and commonly used symbols will also be acceptable provided they are clearly understood, are adequately identified in the specification as filed, and do not create confusion with other symbols used in patent drawings.

NOTES: In general, in lieu of a symbol, a conventional element, combination or circuit may be shown by an appropriately labeled rectangle, square, or circle; abbreviations should not be used unless their meaning is evident and not confusing with the abbreviations used in the suggested symbols. In the electrical symbols an arrow through an element indicates variability thereof, see for example symbols 2, 6, 12; dotted line connection of arrows indicates ganging thereof, see symbol 6; inherent property (as resistance) may be indicated by showing symbol (for resistor) in dotted lines.

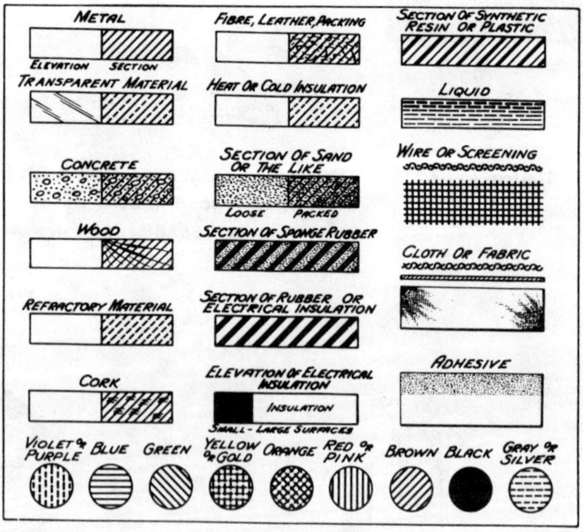

Index